# FROM SUNDIALS
# TO ATOMIC CLOCKS

National Bureau of Standards Atomic Clock NBS-6

# FROM SUNDIALS TO ATOMIC CLOCKS

## Understanding Time and Frequency

James Jespersen
and
Jane Fitz-Randolph

Illustrated by John Robb

DOVER PUBLICATIONS, INC.
New York

Published in Canada by General Publishing Company, Ltd., 30 Lesmill Road, Don Mills, Toronto Ontario.

Published in the United Kingdom by Constable and Company, Ltd.

This Dover edition, first published in 1982, is an unabridged republication of the work first published in 1977, under the same title, as Monograph 155 by the National Bureau of Standards, a division of the U.S. Department of Commerce. Preparation of the original work was supported in part by the 1842nd Electronic Engineering Group, C²/ DCS Division, Air Force Communications. For the Dover edition the color blue of the original edition has been printed as gray.

The comic strip *B.C.* is used by permission of Johnny Hart and Field Enterprises, Inc.

Manufactured in the United States of America
Dover Publications, Inc.
180 Varick Street
New York, N.Y. 10014

**Library of Congress Cataloging in Publication Data**

Jespersen, James.
  From sundials to atomic clocks.

  Reprint. Originally published: Washington: National Bureau of Standards, U.S. Dept. of Commerce: for sale by the Supt. of Docs., U.S. G.P.O., 1977. (NBS monograph; 155)
  Bibliography: p.
  Includes index.
  1. Time.   2. Time measurement.   3. Clocks and watches.
I. Fitz-Randolph, Jane.   II. Robb, John.   III. Title.   IV. Series: NBS monograph; 155.
QC100.U556 no. 155 1982 [QB209] 602′ .1s   81-17320
ISBN 0-486-24265-X          [529]          AACR2

# FOREWORD

Time and its measurement is, simultaneously, very familiar and very mysterious. I suspect we all believe that the readings of our clocks and watches are somehow related to the sun's position. However, as science and technology developed, this relationship has come to be determined by a very complex system involving—just to name a few—astronomers, physicists, electronic engineers, and statisticians. And because time is both actively and precisely coordinated among all of the technologically advanced nations of the world, international organizations are also involved. The standard time-of-day radio broadcasts of all countries are controlled to at least 1/1000 of a second of each other; most time services, in fact, are controlled within a very few millionths of a second!

The National Bureau of Standards (NBS) mounts a major effort in developing and maintaining standards for time and frequency. This effort tends to be highly sophisticated and perhaps even esoteric at times. Of course, most of the publications generated appear in technical journals aimed at specialized, technically sophisticated audiences.

I have long been convinced, however, that it is very important to provide a descriptive book, addressed to a much wider audience, on the subject of time. There are many reasons for this, and I will give two. First, it is—very simply—a fascinating subject. Again, we often have occasion to explain the NBS time program to interested people who do not have a technical background, and such a book would be an efficient and—hopefully—interesting means of informing them. Finally, this book realizes a long-standing personal desire to see a factual and yet understandable book on the subject of time.

James A. Barnes
May 6, 1977

# Contents

## I. THE RIDDLE OF TIME

## II. MAN-MADE CLOCKS AND WATCHES

## III. FINDING AND KEEPING THE TIME

# IV
# THE USES OF TIME

# V
# TIME, SCIENCE, AND TECHNOLOGY

# PREFACE AND ACKNOWLEDGMENTS

This is a book for laymen. It offers an introduction to time, timekeeping, and the uses of time information, especially in the scientific and technical areas. It is impossible to consider time and timekeeping without including historical and philosophical aspects of the subject, but we have merely dabbled in these. We hope historians and philosophers will forgive our shallow coverage of their important contributions to man's understanding of time, and that scientists will be forbearing toward our simplified account of scientific thought on time in the interest of presenting a reasonably complete view in a limited number of pages.

Time is an essential component in most disciplines of science ranging from astronomy to nuclear physics. It is also a practical necessity in managing our everyday lives, in such obvious ways as getting to work on time, and in countless ways that most persons have never realized, as we shall see.

Because of the many associations of time, we have introduced a certain uniformity of language and definition which the specialist will realize is somewhat foreign to his particular field. This compromise seemed necessary in a book directed to the general reader. Today the United States and some parts of the rest of the world are in the process of converting to the metric system of measurement, which we use in this book. We have also used the American definitions of billion and trillion; thus a billion means 1000 million, and a trillion means 1000 billion.

Several sections in this book—the "asides" printed over a light gray background—are included for the reader who wishes to explore a little more fully a particular subject area. These may be safely ignored, however, by the reader who wishes to move on to the next major topic, since understanding the book does not depend upon reading these more "in-depth" sections.

This book could not have been written without the help and support of a number of interested persons. James A. Barnes, Chief of the Time and Frequency Division of the Natural Bureau of Standards, first conceived the idea of writing a book of this kind. He has contributed materially to its contents and has steadfastly supported the authors in their writing endeavors. George Kamas, also of the Time and Frequency Division, played the role of devil's advocate, and for this reason many muddy passages have been cast out or rewritten. Critical and constructive comments from many others also helped to extend and clarify many of the concepts presented. Among these are Roger E. Beehler, Jo Emery, Helmut Hellwig, Sandra Howe, Howland Fowler, Stephen Jarvis, Robert Mahler, David Russell, and Collier Smith—all members of the National Bureau of Standards staff. Thanks go also to John Hall and William Klepczynski of the United States Naval Observatory, and Neil Ashby, Professor of Physics at the University of Colorado. Finally, we thank Joanne Dugan, who has diligently and good naturedly prepared the manuscript in the face of a parade of changes and rewrites.

# DEDICATION

The authors dedicate this book to the many who have contributed to man's understanding of the concept of time, and especially to Andrew James Jespersen, father of one of the authors, who—as a railroad man for almost 40 years—understands better than most the need for accurate time, and who contributed substantially to one of the chapters.

# I
# THE RIDDLE OF TIME

3

# Chapter

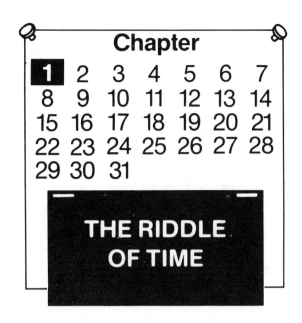

**1** 2 3 4 5 6 7
8 9 10 11 12 13 14
15 16 17 18 19 20 21
22 23 24 25 26 27 28
29 30 31

## THE RIDDLE OF TIME

- It's present everywhere, but occupies no space.
- We can measure it, but we can't see it, touch it, get rid of it, or put it in a container.
- Everyone knows what it is and uses it every day, but no one has been able to define it.
- We can spend it, save it, waste it, or kill it, but we can't destroy it or even change it, and there's never any more or less of it.

All of these statements apply to time. Is it any wonder that scientists like Newton, Descartes, and Einstein spent years studying, thinking about, arguing over, and trying to define time—and still were not satisfied with their answers? Today's scientists have done no better. The riddle of time continues to baffle, perplex, fascinate, and challenge. Pragmatic physicists cannot help becoming philosophical—even metaphysical—when they start pursuing the elusive concepts of time.

Much has been written of a scholarly and philosophical nature. But time plays a vital and practical role in the everyday lives of us all, and it is this practical role which we shall explore in this book.

## THE NATURE OF TIME

**L**ENGTH
**M**ASS
**T**IME
**T**EMPERATURE

Time is a necessary component of many mathematical formulas and physical functions. It is one of several basic quantities from which most physical measurement systems are derived. Others are length, temperature, and mass. Yet time is unlike length or mass or temperature in several ways. For instance—

PAST   NOW   FUTURE

- We can see distance and we feel weight and temperature, but time cannot be apprehended by any of the physical senses. We cannot see, hear, feel, smell, or taste time. We know it only through consciousness, or through observing its effects.
- Time "passes," and it moves in only one direction. We can travel from New York to San Francisco or from San Francisco to New York, moving "forward" in either case. We can weigh the grain produced on an acre of land, beginning at any point, and progressing with any measure "next." But when we think of time, in even the crudest terms, we must always think of it as now, before now, and after now. We cannot *do* anything in either the past or the future—only "now."
- "Now" is constantly changing. We can buy a good one-foot ruler or meter-stick, or a one-gram weight, or even a thermometer, put it away in a drawer or cabinet, and use it whenever we wish. We can forget it between uses—for a day or a week or ten years—and find it as useful when we bring it out as when we put it away. But a "clock"—the "measuring stick" for time—is useful only if it is kept "running." If we put it away in a drawer and forget it, and it "stops," it becomes useless until it is "started" again, and "reset" from information available only from another clock.
- We can write a postcard to a friend and ask him how long his golf clubs are or how much his bowling ball weighs, and the answer he sends on another postcard

gives us useful information. But if we write and ask him what time it is—and he goes to great pains to get an accurate answer, which he writes on another postcard—well, obviously before he writes it down, his information is no longer valid or useful.

DEAR ED —
TO ANSWER YOUR QUESTIONS:
1. MY CLUBS ARE 102 CM. LONG.
2. MY BOWLING BALL WEIGHS 6 KG.
3. THE TIME IS 2:34 P.M.
YOURS,
JERRY

This fleeting and unstable nature of time makes its measurement a much more complex operation than the measurement of length or mass or temperature.

## WHAT IS TIME?

Time is a physical quantity that can be observed and measured with a clock of mechanical, electrical, or other physical nature. Dictionary definitions bring out some interesting points:

- **time**—A nonspatial continuum in which events occur in apparently irreversible succession from past through present to the future. An interval separating two points on this continuum, measured essentially by selecting a regularly recurring event, such as the sunrise, and counting the number of its recurrences during the interval of duration.
  American Heritage Dictionary

- **time**—1. The period during which an action, process, etc. continues; measured or measurable duration . . . . 7. A definite moment, hour, day, or year, as indicated or fixed by the clock or calendar.
  Webster's New Collegiate Dictionary

At least part of the trouble in agreeing on what time is lies in the use of the single word *time* to denote two distinct concepts. The first is *date* or *when* an event happens. The other is *time interval*, or the "length" of time between two events. This distinction is important, and is basic to the problems involved in measuring time. We shall have a great deal to say about it.

## DATE, TIME INTERVAL, AND SYNCHRONIZATION

We obtain the date of an event by counting the number of cycles, and fractions of cycles, of periodic events, such as the sun as it appears in the sky and the earth's movement around the sun, beginning at some agreed-upon starting point. The date of an event might be 13 February 1976, 14h, 35m, 37.27s; *h, m,* and *s* denote hours, minutes, and seconds; the 14th hour, on a 24-hour clock, would be two o'clock in the afternoon.

In the United States literature on navigation, satellite tracking, and geodesy, the word "epoch" is sometimes used in a similar sense to the word "date." But there is considerable ambiguity in the word "epoch," and we prefer the term "date," the precise meaning of which is neither ambiguous nor in conflict with other, more popular uses.

*Time interval* may or may not be associated with a specific date. A person timing the movement of a horse around a race-track, for example, is concerned with the minutes, seconds, and fractions of a second between the moment the horse leaves the gate and the moment it crosses the finish line. The *date* is of interest only if he must have the horse at a particular track at a certain hour on a certain day.

WHEN ?
HOW LONG ?
TOGETHER !

Time interval is of vital importance to *synchronization*, which means literally "timing together." Two military units that expect to be separated by several kilometers may wish to surprise the enemy by attacking at the same moment from opposite sides. So before parting, men from the two units synchronize their watches. Two persons who wish to communicate with each other may not be critically interested in the date of their communication, or even in how long their communication lasts. But unless their equipment is precisely *synchronized*, their messages will be garbled. Many sophisticated electronic communications systems, navigation systems, and proposed aircraft collision-avoidance systems have little concern with accurate dates; but they depend for their very existence on extremely accurate synchronization.

The problem of synchronizing two or more time-measuring devices—getting them to measure time interval accurately and together, very precisely, to the thousandth or millionth of a second —presents a continuing challenge to electronic technology.

## ANCIENT CLOCK WATCHERS

Among the most fascinating remains of many ancient civilizations are their elaborate time-watching devices. Great stone structures like Stonehenge, in Southern England, and the 4,000-year-old passage grave of Newgrange, near Dublin, Ireland, that have challenged anthropologists and archaeologists for centuries, have proved to be observatories for watching the movement of heavenly bodies. Antedating writing within the culture, often by centuries, these crude clocks and calendars were developed by primitive peoples on all parts of our globe. Maya and Aztec cultures developed elaborate calendars in Central and North America. And even today scientists are finding new evidence that stones laid out in formation on our own western plains, such as the Medicine Wheel in northern Wyoming, formerly thought to have only a religious purpose, are actually large clocks. Of course they had religious significance, also, for the cycles of life—the rise and fall of the tides, and the coming and going of the seasons—powers that literally controlled the lives of primitive peoples as they do our own, naturally evoked a sense of mystery and inspired awe and worship.

Astronomy and time—so obviously beyond the influence or control of man, so obviously much older than anything the oldest man in the tribe could remember and as nearly "eternal" as anything the human mind can comprehend—were of great concern to ancient peoples everywhere.

## CLOCKS IN NATURE

The movements of the sun, moon, and stars are easy to observe, and one can hardly escape being conscious of them. But of course there are countless other cycles and rhythms going on around us—and inside of us—all the time. Biologists, botanists, and other life scientists study but do not yet fully understand many "built in" clocks that regulate basic life processes—from periods of animal gestation and ripening of grain to migrations of birds and fish; from the rhythms of heartbeats and breathing to those of the fertile periods of female animals. These scientists talk about "biological time," and have written whole books about it.

Geologists also are aware of great cycles, each one covering thousands or millions of years; they speak and write in terms of "geologic time." Other scientists have identified quite accurately the rate of decay of atoms of various elements—such as carbon 14, for example. So they are able to tell with considerable dependability the age of anything that contains carbon 14. This includes everything that was once alive, such as a piece of wood that could have been a piece of Noah's Ark or the mummified body of a king or a pre-Columbian farmer.

HMM—A MERE 250 MILLION YEARS OLD!

JOHN ROBB

## KEEPING TRACK OF THE SUN AND MOON

Some of the stone structures of the earliest clock watchers were apparently planned for celebrating a single date—Midsummer Day, the day of the Summer solstice, when the time from sunrise to sunset is the longest. It occurs on June 21–22, depending on how near the year is to leap year. For thousands of years, the "clock" that consists of the earth and the sun was sufficient to regulate daily activities. Primitive peoples got up and began their work at sunrise and ceased work at sunset. They rested and ate their main meal about noon. They didn't need to know time any more accurately than this.

But there were other dates and anniversaries of interest; and in many cultures calendars were developed on the basis of the revolutions of the sun, the moon, and the seasons.

If we think of time in terms of cycles of regularly recurring events, then we see that timekeeping is basically a system of counting these cycles. The simplest and most obvious to start with is days—sunrise to sunrise, or more usefully, noon to noon, since the "time" from noon to noon is, for most practical purposes, always the same, whereas the hour of sunrise varies much more with the season.

B.C. BY PERMISSION OF JOHNNY HART AND FIELD ENTERPRISES, INC.

One can count noon to noon with very simple equipment—a stick in the sand or an already existing post or tree, or even one's own shadow. When the shadow points due North—if one is in the northern hemisphere—or when it is the shortest, the sun is at its zenith, and it is noon. By making marks of a permanent or semipermanent nature, or by laying out stones or other objects in a preplanned way, one can keep track of and count days. With slightly more sophisticated equipment, one can count full moons—or months—and the revolutions of the earth around the sun, or years.

It would have been convenient if these cycles had been neatly divisible into one another, but they are not. It takes the earth about 365¼ days to complete its cycle around the sun, and the moon circles the earth about 13 times in 364 days. This gave early astronomers, mathematicians, and calendar makers some thorny problems to work out.

### THINKING BIG AND THINKING SMALL—AN ASIDE ON NUMBERS

Some scientists, such as geologists and paleontologists, think of time in terms of thousands and millions of years. In their vernacular a hundred years more or less is insignificant—too small to recognize or to measure. To other scientists, such as engineers who design sophisticated communication systems and navigation systems, one or two seconds' variation in a year is intolerable because it causes them all sorts of problems. They think in terms of thousandths, millionths, and billionths of a *second*.

The *numbers* they use to express these very small "bits" of time are very large. $\frac{1}{1,000,000}$ of a second, for example, is one *microsecond*. $\frac{1}{1,000,000,000}$ of a second is one *nanosecond*. To keep from having to deal with these cumbersome figures in working out mathematical formulas, they use a kind of shorthand, similar to that used by mathematicians to express a number multiplied by itself several or many times. Instead of writing $2 \times 2 \times 2$, for example, we write $2^3$, and say, "two to the third power." Similarly, instead of writing $\frac{1}{1,000,000}$, or even .000001, scientists who work with very small fractions express a millionth as $10^{-6}$, meaning 0.1 multiplied by itself 6 times. A billionth of a second, or nanosecond, is expressed as $10^{-9}$ second, which is 0.1 multiplied by itself 9 times. They say, "ten to the minus nine power."

A billionth of a second is an almost inconceivably small bit—many thousands of times smaller than the smallest possible "bit" of length or mass that can be measured. We cannot think concretely about how small a nanosecond is; but to give some idea, the impulses that "trigger" the picture lines on the television screen come, just one at a time, at the rate of 15,750 per second. The whole picture "starts over," traveling left to right, one line at a time, the 525 lines on the picture tube, 30 times a second. At this rate it would take 63,000 nanoseconds just to trace out one line.

Millionths and billionths of a second cannot, of course, be measured with a mechanical clock at all. But today's electronic devices can count them accurately and display the count in usable, meaningful terms.

Whether one is counting hours or microseconds, the principle is essentially the same. It's simply a matter of dividing units to be counted into identical, manageable groups. And since time moves steadily in a "straight line" and in only one direction, counting the swings or ticks of the timer—the *frequency* with which they occur—is easier than counting the pellets in a pailful of buckshot, for example. "Bits" of time, whatever their size, follow one another single file, like beads on a string; and whether we're dealing with ten large bits—hours, for example—or 200 billion small bits, such

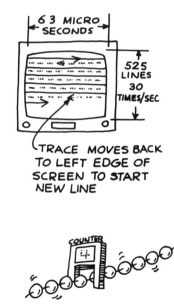

63 MICRO SECONDS

525 LINES 30 TIMES/SEC

TRACE MOVES BACK TO LEFT EDGE OF SCREEN TO START NEW LINE

as microseconds, all we need to do is to count them as they pass through a "gate," and keep track of the count.

The "hour" hand on a clock divides a day evenly into 12 or 24 hours—depending on how the clock face and works are designed. The "minute" hand divides the hour evenly into 60 minutes, and the "second" hand divides the minute evenly into 60 seconds. A "stop watch" has a finer *divider*—a hand that divides the seconds into tenths of a second.

When we have large groups of identical items to count, we often find it faster and more convenient to count by tens, dozens, hundreds, or some other number. Using the same principle, electronic devices can count groups of ticks or oscillations from a frequency source, add them together, and display the results in whatever way one may wish. We may have a device, for example, that counts groups of 9,192,631,770 oscillations of a cesium-beam atomic frequency standard, and sends a special tick each time that number is reached; the result will be very precisely measured one-second intervals between ticks. Or we may want to use much smaller bits—microseconds, perhaps. So we set our electronic divider to group counts into millionths of a second, and to display them on an *oscilloscope*.

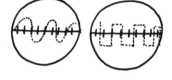

Electronic counters, dividers, and multipliers make it possible for scientists with the necessary equipment to "look at," and to put to hundreds of practical uses, very small bits of time, measured to an accuracy of one or two parts in $10^{11}$; this is about 1 second in 3,000 years.

Days, years, and centuries are, after all, simply units of accumulated nanoseconds, microseconds, and seconds.

# Chapter

| | | | | | | |
|---|---|---|---|---|---|---|
| 1 | **2** | 3 | 4 | 5 | 6 | 7 |
| 8 | 9 | 10 | 11 | 12 | 13 | 14 |
| 15 | 16 | 17 | 18 | 19 | 20 | 21 |
| 22 | 23 | 24 | 25 | 26 | 27 | 28 |
| 29 | 30 | 31 | | | | |

## EVERYTHING SWINGS

The earth swings around the sun, and the moon swings around the earth. The earth "swings" around its own axis. These movements can easily be observed and charted, from almost any spot on earth. The observations were and are useful in keeping track of time, even though early observers did not understand the movements and often were completely wrong about the relationships of heavenly bodies to one another. The "swings" happened with dependable regularity, over countless thousands of years, and therefore enabled observers to predict the seasons, eclipses, and other phenomena with great accuracy, many years in advance.

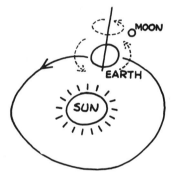

When we observe the earth's swing around its axis, we *see* only a part of that swing, or an arc, from horizon to horizon, as the sun rises and sets. A big breakthrough in timekeeping came when someone realized that another arc—that of a free-swinging pendulum—could be harnessed and adjusted, and its swings counted, to keep track of passing time. The accuracy of the pendulum clock was far superior to any of the many devices that had preceded it—water clocks, hour glasses, candles, and so on. Furthermore, the pendulum made it possible to "chop up" or refine time into much smaller, measurable bits than had ever been possible before; one could measure—quite roughly, to be sure—seconds and even parts of seconds, and this was a great advancement.

The problem of keeping the pendulum swinging regularly was solved at first by a system of cog wheels and an "escapement" that had the effect of giving the pendulum a slight push with each swing, in much the same way that a child's swing is kept in motion by someone pushing it. A weight on a chain kept the

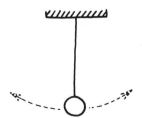

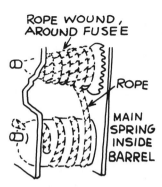

ROPE WOUND,
AROUND FUSE'E

ROPE

MAIN
SPRING
INSIDE
BARREL

AS MAIN SPRING
UNWINDS, LEVERAGE
BETWEEN BARREL AND
FUSÉE CHANGES

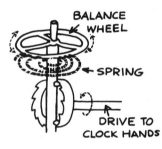

BALANCE
WHEEL

SPRING

DRIVE TO
CLOCK HANDS

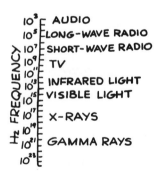

$10^3$ AUDIO
$10^5$ LONG-WAVE RADIO
$10^7$ SHORT-WAVE RADIO
$10^9$ TV
$10^{11}$
$10^{13}$ INFRARED LIGHT
$10^{15}$ VISIBLE LIGHT
$10^{17}$ X-RAYS
$10^{19}$
$10^{21}$ GAMMA RAYS
$10^{23}$

FREQUENCY Hz

escapement lever pushing the pendulum, as it does today in the cuckoo clocks familiar in many homes.

But then someone thought of another way to keep the pendulum swinging—a wound-up spring could supply the needed energy if there were a way to make the "push" from the *partially* wound spring the same as it was from a *tightly* wound spring. The "fusee"—a complicated mechanism that was used for only a brief period—was the answer.

From this it was just one more step to apply a spring and "balance wheel" system directly to the pinions or cogs that turned the hands of the clock, and to eliminate the pendulum. The "swings" were all inside the clock, and this saved space and made it possible to keep clocks moving even when they were moved around or laid on their side.

But some scientists who saw a need for much more precise time measurement than could be achieved by conventional mechanical devices began looking at other things that swing—or vibrate or oscillate—things that swing much faster than the human senses can count. The vibrations of a tuning fork, for instance, which, if it swings at 440 cycles per second, is "A" above "Middle C" on our music scale. The tiny tuning fork in an electric wrist watch, kept swinging by electric impulses from a battery, hums along at 360 vibrations or "cycles" per second.

As alternating-current electricity became generally available at a reliable 60 swings or cycles per second—or 60 hertz (50 in some areas)—it was fairly simple to gear these swings to the clock face of one of the commonest and most dependable time-pieces we have today. For most day-to-day uses, the inexpensive electric wall or desk clock driven by electricity from the local power line keeps "the time" adequately.

But for some users of precise time these common measuring sticks are as clumsy and unsatisfactory as a liter measuring cup would be for a merchant who sells perfume by the dram. These people need something that cuts time up with swings much faster than 60ths or 100ths of a second. The power company itself, to supply electricity at a constant 60 hertz, must be able to measure swings at a much faster rate.

Power companies, telephone companies, radio and television broadcasters, and many other users of precise time have long depended on the swings or vibrations of quartz crystal oscillators, activated by an electric current, to divide time intervals into *megahertz*, or millions of cycles per second. The rate at which the crystal oscillates is determined by the thickness—or thinness—to which it is ground. Typical frequencies are 2.5 or 5 megahertz (MHz)—2½ million or 5 million swings per second.

Incredible as it may seem, it is quite possible to measure swings even much faster than this. What swings faster? Atoms do. One of the properties of each element in the chemistry Periodic Table of Elements is the set of rates at which its atoms swing or resonate. A hydrogen atom, for example, has one of its resonant frequencies at 1,420,405,752 cycles per second, or hertz. A rubidium atom has one at 6,834,682,608 hertz, and a cesium atom at 9,192,631,770 hertz. These are some of the atoms most commonly used in measuring sticks for precise time—the "atomic clocks" maintained by television network master stations, some scientific laboratories, and others. Primary time standards, such as those maintained by the U.S. Naval Observatory or the National Bureau of Standards, are "atomic clocks."

Everything swings, and anything that swings at a constant rate can be used as a standard for measuring time interval.

## GETTING TIME FROM FREQUENCY

The sun as it appears in the sky—or the "apparent sun"—crosses the zenith or highest point in its arc with a "frequency" of once a day, and 365¼ times a year. A metronome ticks off evenly spaced intervals of time to help a musician maintain the time or *tempo* of a composition he is studying. By moving the weight on its pendulum he can slow the metronome's "frequency" or speed it up.

Anything that swings evenly can be used to measure *time interval* simply by counting and keeping track of the number of swings or ticks—provided we know how many swings take place in a recognized unit of time, such as a day, an hour, a minute, or a second. In other words, we can measure time interval if we know the *frequency* of these swings. A man shut up in a dungeon, where he cannot see the sun, could keep a fairly accurate record of passing time by counting his own heartbeats—if he knew how many times his heart beats in one minute—and if he has nothing to do but count and keep track of the number.

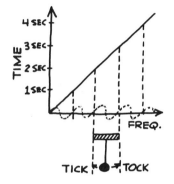

The term *frequency* is commonly used to describe swings too fast to be counted by the human ear, and refers to the number of swings or cycles per second—called hertz (Hz), after Heinrich Hertz, who first demonstrated the existence of radio waves.

If we can count and keep track of the cycles of our swinging device, we can construct a time interval at least as accurate as the device itself—even to millionths or billionths of a second. And by adding these small, identical bits together, we can measure any "length" of time, from a fraction of a second to an hour—or a week or a month or a century.

Of course, the most precise and accurate measuring device in existence cannot tell us the *date*—unless we have a source to tell us when to *start* counting the swings. But if we know this, and if we keep our swinging device "running," we can keep track of both time interval and date by counting the cycles of our device.

## WHAT IS A CLOCK?

Time "keeping" is simply a matter of counting cycles or units of time. A clock is what does the counting. In a more strict definition, a clock also keeps track of its count and displays what it has counted. But in a broad sense, the earth and the sun are a clock—the commonest and most ancient clock we have, and the basis of all other clocks.

When ancient peoples put a stick in the ground to observe the movement of its shadow from sunrise to sunset, it was fairly easy and certainly a natural step to mark off "noon" and other points where the shadow lay at other times of day—in other words, to make a sundial. Sundials can tell the time quite reliably when the sun is shining. But, of course, they are of no use at all when the sun is not shining. So people made mechanical devices called clocks to interpolate or keep track of time between checks with the sun. The sun was a sort of "master clock" that served as a primary time scale by which the man-made, secondary clocks were calibrated and adjusted.

Although some early clocks used the flow of water or sand to measure passing time, the most satisfactory clocks were those that counted the swings of a pendulum or of a balance wheel. Quite recently in the history of timekeeping, men have developed extremely accurate clocks that count the vibrations of a quartz crystal activated by an electric current, or the resonances of atoms of selected elements such as rubidium or cesium. Since "reading" such a clock requires counting millions or billions of cycles per second—in contrast to the relatively slow 24-hour cycle of the earth-sun clock—an atomic clock requires much more sophisticated equipment for making its count. But given the necessary equipment, one can read an atomic clock with much greater ease, in much less time, and with many thousands of times greater precision than he can read the earth-sun clock.

A mechanism that simply swings or ticks—a clockwork with a pendulum, for example, without hands or face—is not, strictly speaking, a clock. The swings or ticks are meaningless, or *ambiguous*, until we are able to count them and until we establish some base from which to *start counting*. In other words, until we hook up "hands" to keep track of the count, and put those hands over a face with numbers that help us count the ticks and oscillations and make note of the accumulated count, we don't have a useful device.

ABOUT
1 CM

METAL CAPSULE CONTAINING QUARTZ CRYSTAL USED IN WRIST WATCHES

The familiar 12-hour clock face is simply a convenient way to keep track of the ticks we wish to count. It serves very well for measuring time interval, in hours, minutes, and seconds, up to a maximum of 12 hours. The less familiar 24-hour clock face serves as a measure of time interval up to 24 hours. But neither will tell us anything about the day, month, or year.

## THE EARTH-SUN CLOCK

As we have observed, the spin of the earth on its axis and its rotation around the sun provide the ingredients for a clock—a very fine clock that we can certainly never get along without. It meets many of the most exacting requirements that the scientific community today makes for an acceptable standard:

- It is *universally available.* Anyone, almost anywhere on earth, can readily read and use it.

  AVAILABILITY

- It is *reliable.* There is no foreseeable possibility that it may stop or "lose" the time, as is possible with all man-made clocks.

  RELIABILITY

- It has great over-all *stability.* On the basis of its time scale, scientists can predict such things as the hour, minute, and second of sunrise and sunset at any part of the globe; eclipses of the sun and moon, and other time-oriented events hundreds or thousands of years in advance.

  STABILITY

In addition, it involves no expense of operation for anyone; there is no possibility of international disagreement as to "whose" sun is the authoritative one, and no responsibility for keeping it running or adjusted.

## B.C.

B.C. BY PERMISSION OF JOHNNY HART AND FIELD ENTERPRISES, INC.

Nevertheless, this ancient and honored timepiece has some serious limitations. As timekeeping devices were improved and became more common—and as the study of the earth and the universe added facts and figures to those established by earlier observers—it became possible to measure quite precisely some of the phenomena that had long been known in a general way, or at least suspected. Among them was the fact that the earth-sun clock is not, by more precise standards, a very stable timepiece.

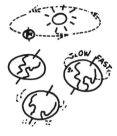

- The earth's orbit around the sun is not a perfect circle but is elliptical; so the earth travels faster when it is nearer the sun than when it is farther away.
- The earth's axis is tilted to the plane containing its orbit around the sun.
- The earth spins at an irregular rate around its axis of rotation.
- It also wobbles on its axis.

For all of these reasons the earth-sun clock is not an accurate clock. The first two facts alone cause the day, as measured by a sun dial, to differ from time, as we reckon it today, by about 15 minutes a day in February and November. These effects are predictable and cause no serious problem, but there are also significant, unpredictable variations.

Gradually, man-made clocks became so much more stable and precise than the earth-sun clock as time scales for measuring short time intervals that solar time had to be "corrected." As mechanical and electrical timepieces became more common and more dependable, as well as easier to use, nearly everyone looked to them for the time and forgot about the earth-sun clock as the master clock. People looked at a clock to see what time the sun rose, instead of looking at the sunrise to see what time it was.

## B.C.

B.C. BY PERMISSION OF JOHNNY HART AND FIELD ENTERPRISES, INC.

## METER-STICKS TO MEASURE TIME

If we have to weigh a truckload of sand, a bathroom scale is of little use. Nor is it of any use for finding out whether a letter will need one postage stamp or two. A meterstick is all right for measuring centimeters—unless we want to measure a thousand or ten thousand meters—but it won't do for measuring accurately the thickness of an eyeglass lens.

Furthermore, if we order a bolt $\frac{5}{16}$ of an inch in diameter and 8-$\frac{3}{16}$ inches long—and our supplier has only a *meter*-stick, he will have to use some arithmetic before he can fill our order. His *scale* is different from ours. Length and mass can be chopped up into any predetermined size bits anyone wishes. Some sizes, of course, are easier to work with than others, and so have come into common use. The important point is that everyone concerned with the measurement agrees on what the *scale* is to be. Otherwise a liter of tomato juice measured by the juice processor's scale might be quite different from the liter of gasoline measured by the oil company's scale.

Time, too, is measured by a scale. For practical reasons, the already existing scale, set by the spinning of the earth on its axis and the rotation of the earth around the sun, provides the basic scale from which others have been derived.

## WHAT IS A STANDARD?

We have noted that the important thing about measurement is that there be general agreement on exactly what the *scale* is to be, and how the basic *unit* of that scale is to be defined. In other words, there must be agreement upon the *standard* against which all other measurements and calculations will be compared. In the United States the standard unit for measuring length is the meter. The basic unit for measurement of mass is the gram.

The basic unit for measuring time is the second. The second multiplied evenly by 60 gives us minutes, or by 3600 gives us hours. The length of days, and even years, is measured by the basic unit of time, the second. Time intervals of less than a second are measured in 10ths, 100ths, 1000ths—on down to billionths of a second.

Each basic unit of measurement is very exactly and explicitly defined by international agreement; and then each nation directs a government agency to make *standard units* available to anyone who wants them. In our country, the National Bureau of Standards (NBS), a part of the Department of Commerce with headquarters in Gaithersburg, Maryland, provides the primary standard references for ultimate calibration of the many standard weights and measures needed for checking scales in drug and grocery stores, the meters that measure the gasoline we pump into our cars, the octane of that gasoline, the purity of the gold in our jewelry or dental repairs, the strength of the steel used in automobile parts and children's tricycles, and countless other things that have to do with the safety, efficiency, and comfort of our everyday lives.

The National Bureau of Standards is also responsible for making the second—the standard unit of time interval—available

to many thousands of time users everywhere—not only throughout the land, but to ships at sea, planes in the air, and even vehicles in outer space. This is a tremendous challenge, for the standard second, unlike the standard meter or kilogram, cannot be sent in an envelope or box and put on a shelf for future reference, but must be supplied constantly, on a ceaseless basis, from moment to moment—and even counted upon to give the date.

### HOW TIME TELLS US WHERE IN THE WORLD WE ARE

One of the earliest, most vital, and universal needs for precise time information was—and still is—as a basis for place location. Navigators of ships at sea, planes in the air, and even small pleasure boats and private aircraft depend constantly and continuously on time information to find out where they are and to chart their course. Many people know this, in a general way, but few understand how it works.

Primitive man discovered long ago that the sun and stars could aid him in his travels, especially on water where there are no familiar "signposts." Early explorers and adventurers in the northern hemisphere were particularly fortunate in having a "pole star," the North Star, that appeared to be suspended in the northern night sky; it did not rotate or change its position with respect to Earth as the other stars did.

These early travelers also noticed that as they traveled northward, the North Star gradually appeared higher and higher in the sky, until it was directly overhead at the North Pole. By measuring the elevation of the North Star above the horizon, then, a navigator could determine his distance from the North Pole—and conversely, his distance from the equator. An instrument called a *sextant* helped him measure this elevation very accurately. The measurement is usually indicated in *degrees of latitude*, ranging from 0 degrees latitude at the equator to 90 degrees of latitude at the North Pole.

Measuring distance and charting a course east or west, however, presented a more complex problem because of the earth's spin. But the problem also provides the key to its solution.

For measurements in the east-west direction, the earth's surface has been divided into lines of *longitude,* or meridians; one complete circuit around the earth equals 360 degrees of longitude, and all longitude lines intersect at the North and South Poles. By international agreement, the line of longitude that runs through Greenwich, England, has been labeled the zero meridian; and longitude is measured east and west from this meridian to the point where the measurements meet at 180 degrees, on the opposite side of the earth from the zero meridian.

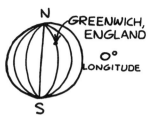

At any point on earth, the sun travels across the sky from east to west at the rate of 15 degrees in one hour, or one degree in four minutes. So if a navigator has a very accurate *clock* aboard his ship—one that can tell him very accurately the time at Greenwich or the zero median—he can easily figure his longitude. He simply gets the time *where he is* from the sun. For every four minutes that his clock, showing Greenwich time, differs from the time determined locally from the sun, he is one degree of longitude away from Greenwich.

At night he can get his position by observing the location of two or more stars. The method is similar to obtaining latitude from the North Star. The difference is that whereas the North Star appears suspended in the sky, the other stars appear to move in circular paths around the North Star. Because of this, the navigator must know the *time* in order to find out where he is. If he does not know the time, he can read his location with respect to the *stars,* as they "move" around the North Star, but he has no way at all to tell where he is *on earth*! His navigation charts tell him the positions of the stars at any given *time* at every season of the year; so if he knows the time, he can find out where he is simply by referring to two or more stars, and reading his charts.

The principle of the method is shown in the illustration. For every star in the sky there is a point on the surface of the earth where the star appears directly overhead. This is Point A for Star #1 and Point B for Star #2 in the illustration. The traveler at Point 0 sees Star #1 at some angle from the overhead position. But as the illustration shows, all travelers standing on the black circle will see Star #1 at this same angle. By observing Star #2, the traveler will put himself on another circle of points, the gray circle; so his location will be at one of the two intersection points of the gray and the black circles.

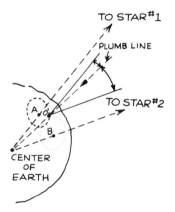

He can look at a third star to choose the correct intersection point; or, as is more usually the case, he has at least some idea of his location, so that he can pick the correct intersection point without further observation.

The theory is simple. The big problem was that until about 200 years ago, no one was able to make a clock that could keep time accurately at sea.

## BUILDING A CLOCK THAT WOULDN'T GET SEASICK

During the centuries of exploration of the world that lay thousands of miles across uncharted oceans, the need for improved navigation instruments became critical. Ship building improved, and larger, stronger vessels made ocean trade—as well as ocean warfare—increasingly important. But too often ships laden with priceless merchandise were lost at sea, driven off course by storms, with the crew unable to find out where they were or to chart a course to a safe harbor.

Navigators had long been able to read their latitude north of the equator by measuring the angle formed by the horizon and the North Star. But east-west navigation was almost entirely a matter of "dead reckoning." If only they had a *clock* aboard that could tell them the time at Greenwich, England, then it would be easy to find their position east or west of the zero meridian.

It was this crucial need for accurate, dependable clocks aboard ships that pushed inventors into developing better and better timepieces. The pendulum clock had been a real breakthrough, and an enormous improvement over any timekeeping device made before it. But it was no use at all at sea. The rolling and pitching of the ship made the pendulum inoperative.

In 1713 the British government offered an award of £20,000 to anyone who could build a chronometer that would serve to determine longitude to within ½ degree. Among the many craftsmen who sought to win this handsome award was an English clock maker named John Harrison, who spent more than 40 years trying to meet the specifications. Each model became a bit more promising as he found new ways to cope with the rolling sea, temperature changes that caused intolerable expansion and contraction of delicate metal springs, and salt spray that corroded everything aboard ship.

ROLLING SEA
TEMPERATURE
SALT SPRAY

When finally he came up with a chronometer that he considered nearly perfect, the men of the government commission were so afraid that it might be lost at sea that they suspended testing it until Harrison had built a second unit identical with the first, to provide a pattern. Finally, in 1761 Harrison's son William was sent on a voyage to Jamaica to test the instrument. In spite of a severe storm that lasted for days and drove the ship far off course, the chronometer proved to be amazingly accurate, losing less than 1 minute over a period of many months and making it possible for William to determine his longitude at sea within 18 minutes of arc, or less than ⅓ of one degree. Harrison claimed the £20,000 award, part of which he had already received, and the remainder was paid to him in various amounts over the next two years—just three years before his death.

For more than half a century after Harrison's chronometer was accepted, an instrument of similar design—each one built entirely by hand, of course, by a skilled horologist—was an extremely valuable and valued piece of equipment—one of the most vital items aboard a ship. It needed very careful tending, and the one whose duty it was to tend it had a serious responsibility.

Today there may be almost as many wrist watches as crew aboard an ocean-going ship—many of them as accurate and dependable as Harrison's prized chronometer. But the ship's chronometer, built on essentially the same basic principles as Harrison's instrument, is still a most vital piece of the ship's elaborate complement of navigation instruments.

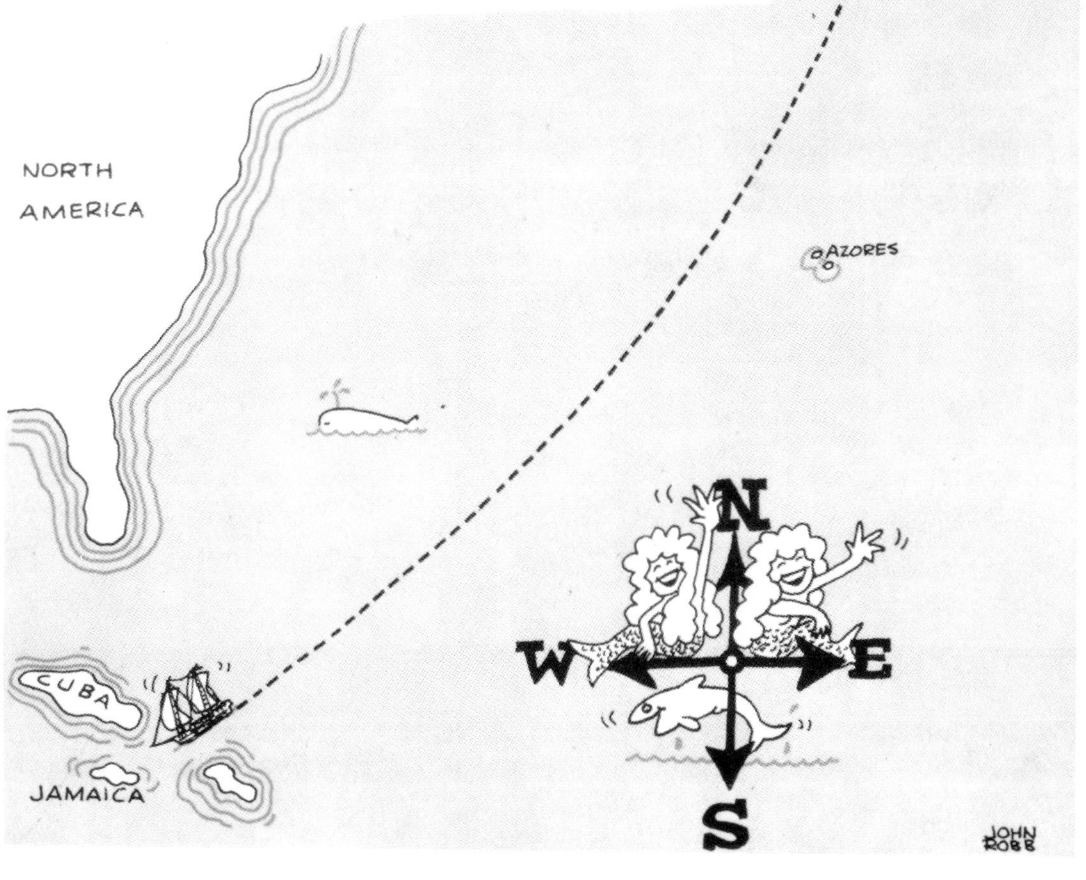

LONDON

NORTH
AMERICA

AZORES

N

W      E

S

CUBA

JAMAICA

JOHN
ROBB

## II
## MAN-MADE CLOCKS AND WATCHES

# Chapter

| 1 | 2 | **3** | 4 | 5 | 6 | 7 |
| 8 | 9 | 10 | 11 | 12 | 13 | 14 |
| 15 | 16 | 17 | 18 | 19 | 20 | 21 |
| 22 | 23 | 24 | 25 | 26 | 27 | 28 |
| 29 | 30 | 31 | | | | |

## EARLY MAN-MADE CLOCKS

Three young boys, lured by the fine weather on a warm spring day, decided to skip school in the afternoon. The problem was knowing when to come home, so that their mothers would think they were merely returning from school. One of the boys had an old alarm clock that would no longer run, and they quickly devised a scheme: The boy with the clock set it by a clock at home when he left after lunch at 12:45. After they met they would take turns as timekeeper, counting to 60 and moving the minute hand ahead one minute at a time!

Almost immediately two of the boys got into an argument over the rate at which the third was counting, and *he* stopped counting to defend his own judgment. They had "lost" the time—crude as their system was—before their adventure was begun, and spent most of their afternoon alternately accusing one another and trying to estimate how much time their lapses in counting had consumed.

"Losing" the time is a constant problem even for timekeepers much more sophisticated than the boys with their old alarm clock. And regulating the clock so that it will "keep" time accurately, even with high-quality equipment, presents even greater challenges. We have already discussed some of these difficulties, in comparison with the relatively simple keeping of a device for measuring length or mass, for example. We've talked about what a clock is, and have mentioned briefly several different kinds of clocks. Now let's look more specifically at the components common to all clocks, and the features that distinguish one kind of clock from another.

EARLY CHINESE
WATER CLOCK

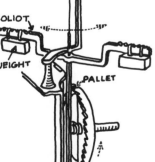

## SAND AND WATER CLOCKS

The earliest clocks that have survived to the present time were built in Egypt. The Egyptians constructed both sundials and water clocks. The water clock in its simplest form consisted of an alabaster bowl, wide at the top and narrow at the bottom, marked on the inside with horizontal "hour" marks. The bowl was filled with water that leaked out through a small hole in the bottom. The clock kept fairly uniform time because more water ran out between hour marks when the bowl was full than when it was nearly empty and the water leaked out more slowly.

The Greeks and Romans continued to rely on water and sand clocks, and it was not until sometime between the 8th and 11th centuries A.D. that the Chinese constructed a clock that had some of the characteristics of later "mechanical" clocks. The Chinese clock was still basically a water clock, but the falling water powered a water wheel with small cups arranged at equal intervals around its rim. As a cup filled with water it became heavy enough to trip a lever that allowed the next cup to move into place; and thus the wheel revolved in steps, keeping track of the time.

Many variations of the Chinese water clock were constructed, and it had become so popular by the early 13th century that a special guild for its makers existed in Germany. But aside from the fact that the clock did not keep very good time, it also tended to freeze in the western European winter.

The sand clocks introduced in the 14th century avoided the freezing problem. But because of the weight of the sand, they were limited to measuring short intervals of time. One of the chief uses of the hour glass was on ships. Sailors threw overboard a log with a long rope attached to it. As the rope played out into the water, they counted knots tied into it at equal intervals, for a specified period of time as determined by the sand clock. This gave them a crude estimate of the speed—or "knots"—at which the ship was moving.

## MECHANICAL CLOCKS

The first mechanical clock was built probably sometime in the 14th century. It was powered by a weight attached to a cord wrapped around a cylinder. The cylinder in turn was connected to a notched wheel, the *crown wheel*. The crown wheel was constrained to rotate in steps by a vertical mechanism called a *verge escapement*, which was topped by a horizontal iron bar, the *foliot*, with movable weights at each end. The foliot was pushed first in one direction and then the other by the crown wheel, the teeth of which engaged small metal extensions called *pallets* at the top and bottom of the crown wheel. Each time the foliot moved back and forth, one tooth of the crown wheel was allowed to escape. The rate of the clock was adjusted by moving the weights in or out along the foliot.

Since the clock kept time only to about 15 minutes a day, it did not need a minute hand. No two clocks kept the same time because the period was very dependent upon the friction between parts, the weight that drove the clock, and the exact mechanical arrangement of the parts of the clock. Later in the 15th century the weight was replaced by a spring in some clocks; but this was also unsatisfactory because the driving force of the spring diminished as the spring unwound.

### The Pendulum Clock

As long as the period of a clock depended primarily upon a number of complicated factors such as friction between the parts, the force of the driving weight or spring, and the skill of the craftsman who made it, clock production was a chancy affair, with no two clocks showing the same time, let alone keeping accurate time. What was needed was some sort of periodic device whose frequency was essentially a property of the device itself and did not depend primarily on a number of external factors.

A pendulum is such a device. Galileo is credited with first realizing that the pendulum could be the frequency-determining device for a clock. As far as Galileo could tell, the period of the pendulum depended upon its length and not on the magnitude of the swing or the weight of the mass at the end of the string. Later work showed that the period does depend slightly upon the magnitude of the swing, but this correction is small as long as the magnitude of the swing is small.

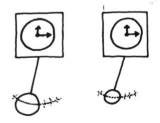

Apparently Galileo did not get around to building a pendulum clock before he died in 1642, leaving this application of the principle to the Dutch scientist Christian Huygens, who built his first clock in 1656. Huygens' clock was accurate to ten seconds a day—a dramatic improvement over the "foliot" clock.

### The Balance Wheel Clock

At the same time that Huygens was developing his pendulum clock, the English scientist Robert Hooke was experimenting with the idea of using a straight metal spring to regulate the frequency

of a clock. But it was Huygens who, in 1675, first successfully built a spring-controlled clock. He used a spiral spring, the derivative of which—the "hair spring"—is still employed in watches today. We have already told the story of John Harrison, the Englishman who built a clock that made navigation practical. The rhythm of Harrison's chronometer was maintained by the regular coiling and uncoiling of a spring. One of Harrison's chronometers gained only 54 seconds during a five-month voyage to Jamaica, or about one-third second per day.

### Further Refinements

The introduction of the pendulum was a giant step in the history of keeping time. ·But nothing material is perfect. Galileo correctly noted that the period of the pendulum depends upon its length; so the search was on for ways to overcome the expansion and contraction of the length of the pendulum caused by changes in temperature. Experiments with different materials and combinations of metals greatly improved the situation.

But another troublesome problem was that as the pendulum swings back and forth it encounters friction caused by air drag, and the amount of drag changes with atmospheric pressure. This problem can be overcome by putting the pendulum in a vacuum chamber; but even with this refinement there are still tiny amounts of friction that can never be completely overcome. So it is always necessary to recharge the pendulum occasionally with energy, but the very process of recharging slightly alters the period of the pendulum.

Attempts to overcome all of these difficulties finally led to a clock that had two pendulums—the "free" pendulum and the "slave" pendulum. The free pendulum was the frequency-keeping device, and the slave pendulum controlled the release of energy to the free pendulum and counted its swings. This type of clock kept time to a few seconds in five years.

SCHEMATIC DRAWING OF AN EARLY TWO-PENDULUM CLOCK. THE SLAVE PENDULUM TIMES THE RELEASE OF ENERGY VIA AN ELECTRIC CIRCUIT TO THE FREE PENDULUM, THUS AVOIDING A DIRECT MECHANICAL CONNECTION BETWEEN THE FREE AND SLAVE PENDULUM.

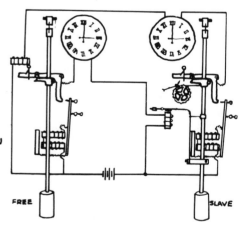

FREE

SLAVE

## THE SEARCH FOR EVEN BETTER CLOCKS

If we are to build a better clock, we need to know more about how a clock's major components contribute to its performance. We need to understand "what makes it tick." So before we begin the discussion of today's advanced atomic clocks, let's digress for a few pages to talk about the basic components of all clocks and how their performance is measured.

From our previous discussions we can identify three main features of all clocks:

- We must have some device that will produce a "periodic phenomenon." We shall call this device a *resonator*.
- We must sustain the periodic motion by feeding energy to the resonator. We shall call the resonator and the energy source, taken together, an *oscillator*.
- We need some means for counting, accumulating, and displaying the ticks or swings of our oscillator—the hands on the clock, for example.

POWER ← RESONATOR → DISPLAY

All clocks have these three components in common.

FAUCET **A** IS TURNED ON SO WATER FLOWS THROUGH HOSE **B** AND DRIPS (**C**) AT THE RATE OF 6 DROPS PER MINUTE INTO BUCKET **D** (ATTACHED TO BOARD **E**). BUCKET PIVOTS AT **F** WHEN WEIGHT OF 360TH DROP FALLS CAUSING BONE **G** IN DOG DISH **H** (ATTACHED TO BOARD **E**) TO CATAPULT AGAINST CHINESE GONG **I** WHICH WAKES UP SLEEPING DOG **J** WHO RETRIEVES BONE **G** AND PLACES IT IN DOG DISH **H**. DOG **J** THEN GOES BACK TO SLEEP.

JOHN ROBB

## HOUR CLOCK

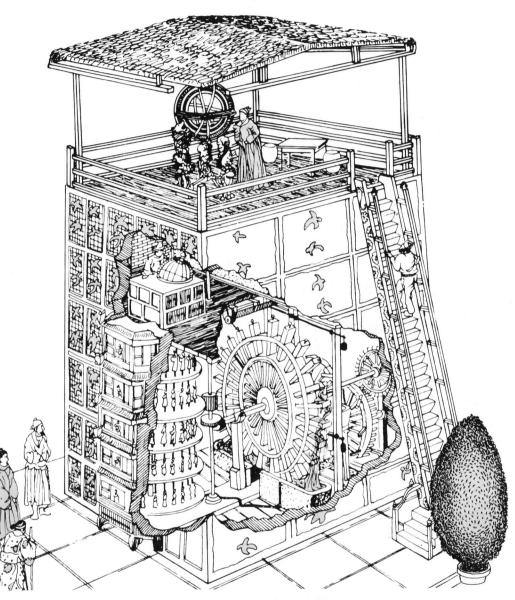

14TH CENTURY CHINESE WATER CLOCK

# Chapter

| 1 | 2 | 3 | **4** | 5 | 6 | 7 |
| 8 | 9 | 10 | 11 | 12 | 13 | 14 |
| 15 | 16 | 17 | 18 | 19 | 20 | 21 |
| 22 | 23 | 24 | 25 | 26 | 27 | 28 |
| 29 | 30 | 31 |

## "Q" IS FOR QUALITY

An ideal resonator would be one that, given a single initial push, would run forever. But of course this is not possible in nature; because of friction everything eventually "runs down." A swinging pendulum comes to a standstill unless we keep replenishing its energy to keep it going.

Some resonators, however, are better than others, and it is useful to have some way of judging the relative merit of various resonators in terms of how many swings they make, given an initial push. One such measure is called the "Quality Factor," or "$Q$." $Q$ is the number of swings a resonator makes until its energy diminishes to a few percent of the energy imparted with the initial push. If there is considerable friction, the resonator will die down rapidly; so resonators with a lot of friction have a low $Q$, and vice versa. A typical mechanical watch might have a $Q$ of 100, whereas scientific clocks have $Q$'s in the millions.

$Q$ = QUALITY FACTOR

One of the obvious advantages of a high-$Q$ resonator is that we don't have to perturb its natural or *resonant* frequency very often with injections of energy. But there is another important advantage. A high-$Q$ resonator won't oscillate at all unless it is swinging at or near its natural frequency. This feature is closely related to the *accuracy* and *stability* of the resonator. A resonator that won't run at all unless it is near its natural frequency is potentially more *accurate* than one that could run at a number of different frequencies. And similarly, if there is a wide range of frequencies over which the resonator can operate, it may wander around within the allowed frequency range, and so will not be very *stable*.

## THE RESONANCE CURVE

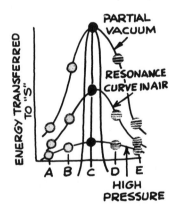

To understand these implications better we shall consider the results of some experiments with the device shown in the sketch. This is simply a wooden frame enclosing a pendulum. At the top of the pendulum is a round wooden stick to which we can attach the pendulums of various lengths shown in the sketch.

Let's begin by attaching pendulum C to the stick and giving it a push. A little bit of the swinging motion of C will be transmitted to the pendulum in the frame, which we shall call S. Since S and C have the same length, their *resonant frequencies* will be the same. This means that S and C will swing with the same frequency, so the swinging energy of C can easily be transferred to S. The situation is similar to pushing someone on a playground swing with the correct timing; we are pushing always *with* the swings, and never working against them.

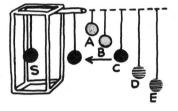

After a certain interval of time we measure the amplitude of the swings of S, which is also a measure of the energy that has been transferred from C to S. The sketch shows this measurement graphically; the black dot in the middle of the graph gives the result of this part of our experiment.

Now let's repeat the experiment, but this time we'll attach pendulum D to the stick. D is slightly longer than S, so its period will be slightly longer. This means that D will be pushing S in the direction it "wants" to swing part of the time, but at other times S will want to reverse its direction before D is ready to reverse. The net result, as shown on our graph by the gray dot above D, is that D cannot transfer energy as easily as could C.

Similarly, if we repeat the experiment with pendulum E attached to the stick, there will be even less transference of energy to S because of E's even greater length. And as we might anticipate, we obtain similar diminishing in energy transfer as we attach pendulums of successively *lesser* length than S. In these cases, however, S will want to reverse its direction at a rate *less* than that of the shorter pendulums.

The results of all our measurements are shown by the second, or middle, curve on our graph; and from now on we shall refer to such curves as the *resonance curve*.

We want to repeat these measurements two more times; first, with the frame in a pressurized chamber, and second with the frame in a partial vacuum. The results of these experiments are shown on the graph. As we might expect, the resonance curve obtained by doing the experiment under pressure is much flatter than that of the experiment performed simply in air. This is true because, at high pressure, the molecules of air are more congested, and so the pendulum experiences a greater frictional loss because of air drag. Similarly, when we repeat the experiment in a partial vacuum, we obtain a sharper, more peaked resonance curve because of reduced air drag.

These experiments point to an important fact for clock builders: the smaller the friction or energy loss, the sharper and more peaked the resonance curve. $Q$, we recall, is related to frictional losses; the lower the friction for a given resonator, the higher the $Q$. Thus we can say that high-$Q$ resonators have sharply peaked resonance curves, and that low-$Q$ resonators have low, flat resonance curves. Or to put it a little differently, the longer it takes a resonator to die down, or "decay," given an initial push, the sharper its resonance curve.

### ENERGY BUILD-UP AND THE RESONANCE CURVE—AN ASIDE ON Q

Why do resonators with a long "decay" time resist running at frequencies other than their natural frequency? A pendulum with a high $Q$ may swing for many minutes, or even hours, from just a single push, whereas a very low-$Q$ pendulum—such as one suspended in honey—may hardly make it through even one swing after an initial push; it would need a new push for every swing, and would never accumulate enough energy to make more than the single swing.

But if we push the high-$Q$ pendulum occasionally in step with its own natural rhythm or frequency, it accumulates or stores up the energy imparted by these pushes. Thus the energy of the pendulum or oscillator may eventually greatly *exceed* the energy imparted by a single push or injection. We can observe this fact by watching someone jumping on a trampoline. As the jumper matches his muscular rhythm to that of his contact with the trampoline, it tosses him higher with each jump; he stores up the energy he puts into it with each jump.

The same principle governs a person swinging on a playground swing. He "pumps up" by adding an extra shot of energy at just the right moment in the swing's natural rhythm or frequency. When he does this, the swing carries over extra energy from his pushes. The rhythm of the swing becomes so strong, in fact, that it can resist or "kick back" at the energy source if it applies energy at the wrong time—as anyone who has pushed someone else in a swing well knows!

In just such a way, a high-$Q$ resonator can accumulate or pile up the energy it receives from its "pusher," or oscillator. But a low-$Q$ resonator cannot accumulate any appreciable amount of energy; instead, the energy will constantly "leak out" at about the same rate it is being supplied, because of friction. Even though we feed the resonator with energy at its natural frequency, the *amplitude* will never build up. On the other hand, if we replenish its energy at a rate other than the natural frequency, the resonator won't have accumulated any appreciable amount of energy at its natural frequency to resist pushes at the wrong rate.

Thus the shape of the resonance curve is determined by the $Q$ of the resonator that is being pushed or driven by some other oscillator, and the transferral of energy from the oscillator to the driven resonator depends upon the similarity between the *natural* frequency of the resonator and the frequency of the oscillator.

## THE RESONANCE CURVE AND THE DECAY TIME

We have already observed that resonators with a high $Q$ or long decay time have a sharp resonance curve. Careful mathematical analysis shows that there is an exact relation between the decay time and the sharpness of the resonance curve, if the sharpness is measured in a particular way. This measurement is simply the *width* of the resonance curve, in hertz (Hz), at the point where the *height* of the curve is half its maximum value.

To illustrate this principle we have redrawn the two resonance curves for our resonator in a pressure chamber and in a partial vacuum. At the half-energy point of the high-pressure curve, the width is about 10 Hz, whereas for the partial vacuum curve the width is about 1 Hz at the half-energy point. With this measurement of width the mathematical analysis shows that the width of the resonance curve at the half-energy point is just one over the decay time of the resonator. As an example, let's suppose it takes a particular resonator 10 seconds to die down or decay. Then the width of its resonance curve at the half-energy point is one over 10 seconds, or 0.1 Hz.

We can think of the width of the curve at the half-energy point as indicating how close the pushes of the driving oscillator must be to the natural frequency of the resonator before it will respond with any appreciable vibration.

## ACCURACY, STABILITY, AND Q

Two very important concepts to clockmakers are *accuracy* and *stability*; and, as we suggested earlier, both are closely related to Q.

We can understand the distinction between accuracy and stability more clearly by considering a machine that fills bottles with a soft drink. If we study the machine we might discover that it fills each bottle with almost exactly the same amount of liquid, to better than $\frac{1}{10}$ of an ounce. We would say the filling *stability* of the machine is quite good. But we might also discover that each bottle is being filled to only half capacity—but very precisely to half capacity from one bottle to the next. We would then characterize the machine as having good stability but poor *accuracy*.

However, the situation might be reversed. We might notice that a different machine was filling some bottles with an ounce or so of extra liquid, and others with an ounce or so less than actually desired, but that *on the average* the correct amount of liquid was being used. We could characterize this machine as having poor *stability* but good *accuracy* over one day's operation.

Some resonators have good stability, others have good accuracy; and the best, for clockmakers, must have both.

### High Q and Accuracy

We have seen that high-Q resonators have long decay times and therefore sharp, narrow resonance curves—which also implies that the resonator won't respond very well to pushes unless they are at a rate very near its natural or resonant frequency. Or to put it differently, a clock with a high-Q resonator essentially won't run at all unless it's running at its resonant frequency.

Today the second of time is defined in terms of a particular resonant frequency of the cesium atom. So if we can build a resonator whose natural frequency is the particular natural frequency of the cesium atom—and furthermore, if this resonator has an extremely high Q—then we have a device that will *accurately* generate the second of time according to the definition of the second.

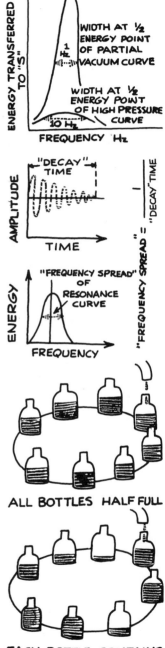

ALL BOTTLES HALF FULL

EACH BOTTLE CONTAINS DIFFERENT AMOUNT

## High Q and Stability

We saw that a low-stability bottle-filling machine is one that does not reliably fill each bottle with the same amount of liquid. And further, that good stability does not necessarily mean high accuracy. A resonator with a high-$Q$, narrow-resonance curve will have good stability because the narrow resonance curve constrains the oscillator to run always at a frequency near the natural frequency of the resonator. We could, however, have a resonator with good stability but whose resonance frequency is not according to the definition of the second—which is the particular natural frequency of the cesium atom. A clock built from such a resonator would have good stability but poor accuracy.

## Waiting to Find the Time

In our discussion of the bottle-filling machine we considered the case of a machine that did not fill each bottle with the desired amount, but that on the average over a day's operation used the correct amount of liquid. We said such a machine had poor stability but good accuracy averaged over a day. The same can be said of clocks. A given clock's frequency may "wander around" within its resonance curve so that for a given measurement the frequency may be in error. But if we average many such measurements over a period of time—or average the time shown by many different clocks at the same time—we can achieve greater accuracy—assuming, of course, that the resonator's natural frequency is the correct frequency.

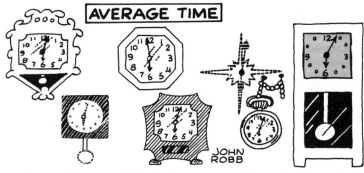

AVERAGE TIME

JOHN ROBB

It would appear that clock error could be made as small as desired if enough measurements were averaged over a long period of time. But experience shows that this is not true. As we first begin to average the measurements, we find that the fluctuations in frequency do decrease; but then beyond some point the fluctuations no longer decrease with averaging, but remain rather constant. And finally, with more measurements considered in the averaging, the frequency stability begins to grow worse again.

The reasons that averaging does not improve clock performance beyond a certain point are not entirely understood. The phenomenon is referred to as "flicker" noise and has been observed in

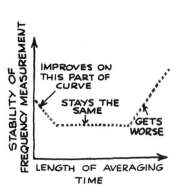

STABILITY OF FREQUENCY MEASUREMENT

IMPROVES ON THIS PART OF CURVE

STAYS THE SAME

GETS WORSE

LENGTH OF AVERAGING TIME

other electronic devices—and interestingly enough, even in the fluctuations of the height of the Nile River.

## PUSHING Q TO THE LIMIT

One may wonder whether there is any limit to how great $Q$ may be. Or in other words, whether clocks of arbitrarily high accuracy and stability can be constructed. It would appear that there is no fundamental reason why $Q$ cannot be arbitrarily high, although there are some practical considerations that have to be accounted for, especially when $Q$ is very high. We shall consider this question in more detail later, when we discuss resonators based upon atomic phenomena; but we can make some general comments here.

Extremely high $Q$ means that the resonance curve is extremely narrow, and this fact dictates that the resonator will not resonate unless it is being driven by a frequency very near its own resonant frequency. But how are we to generate such a driving signal with the required frequency?

The solution is somewhat similar to tuning in a radio station —or tuning one stringed instrument to another. We let the frequency of the driving signal change until we get the maximum response from the high-$Q$ resonator. Once the maximum response is achieved, we attempt to maintain the driving signal at the frequency that produced this response. In actual practice this is done by using a "feedback" system of the kind shown in the sketch.

We have a box that contains our high-$Q$ resonator, and we feed a signal to it from our other oscillator, whose output frequency can be varied. If the signal frequency from the oscillator is near the resonant frequency of the high-$Q$ resonator, it will have considerable response and will produce an output signal voltage proportional to its degree of response. This signal is fed back to the oscillator in such a way that it controls the output frequency of the resonator. This system will search for that frequency from the oscillator which produces the maximum response from the high-$Q$ resonator, and then will attempt to maintain that frequency.

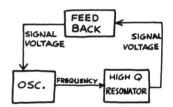

On page 41, where we discuss resonators based upon atomic phenomena, we shall consider feedback again. With a fair notion of what "$Q$" is all about and of how it describes the potential stability and accuracy of a clock, we are in a position to understand a number of other concepts introduced later in this book.

| TYPE | Q |
|---|---|
| INEXPENSIVE BALANCE WHEEL WATCH | 1000 |
| TUNING FORK WATCH | 2000 |
| QUARTZ CLOCK | $10^5 - 10^6$ |
| RUBIDIUM CLOCK | $10^6$ |
| CESIUM CLOCK | $10^7 - 10^8$ |
| HYDROGEN MASER CLOCK | $10^9$ |

# Chapter

| 1 | 2 | 3 | 4 | **5** | 6 | 7 |
| 8 | 9 | 10 | 11 | 12 | 13 | 14 |
| 15 | 16 | 17 | 18 | 19 | 20 | 21 |
| 22 | 23 | 24 | 25 | 26 | 27 | 28 |
| 29 | 30 | 31 | | | | |

# BUILDING EVEN BETTER CLOCKS

The two-pendulum clock—developed in 1921 by William Hamilton Shortt—squeezed just about the last ounce of perfection out of mechanical clocks. If significant gains were to be made, a new approach was needed. As we shall see, new approaches became available because of man's increased understanding of nature—particularly in the realms of electricity, magnetism, and the atomic structure of matter. In one sense, however, the new approaches were undertaken within the framework of the old principles. The heart of the clock is today, as it was 200 year ago, some vibrating device with a period as uniform as possible.

Furthermore, the periodic phenomena today, as before, involve the conversion of energy, to and fro, between two different *forms*. In the swinging pendulum we have energy being transferred back and forth repeatedly from the maximum energy of motion—*kinetic* energy—at the bottom of the swing, to energy stored in the pull of the earth's gravity—or *potential* energy—at the top of the swing. If the energy does not "leak out" because of friction, the pendulum swings back and forth forever, continually exchanging its energy between the two forms.

Energy appears in many forms—kinetic, potential, heat, chemical, light ray, electrical, and magnetic fields. In this discussion we shall be particularly interested in the way energy is transferred between atoms and surrounding fields of radio and light waves. And we shall see that resonators based on such phenomena have achieved $Q$'s in the hundreds of *millions*.

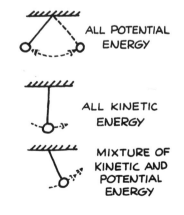

ALL POTENTIAL ENERGY

ALL KINETIC ENERGY

MIXTURE OF KINETIC AND POTENTIAL ENERGY

### THE QUARTZ CLOCK—Q = 10⁵ — 10⁶

The first big step in a new direction was taken by the American scientist Dr. Warren A. Marrison with the development of the quartz crystal clock in 1929. The resonator of this clock is based upon the so-called "piezoelectric effect." In a sense even the quartz crystal clock is actually a mechanical clock because a small piece of quartz crystal vibrates when an alternating electric voltage is applied to it. Or, conversely, if the crystal is made to vibrate it will generate an oscillatory voltage. These two phenomena together are the piezoelectric effect. The internal friction of the quartz crystal is so very low that the $Q$ may range from 100,000 to 1,000,000. It is no wonder that the quartz resonator brought such dramatic gains to the art of building clocks.

The resonant frequency of the crystal depends in a complicated way on how the crystal is cut, the size of the crystal, and the particular resonant frequency that is excited in the crystal by the driving electric voltage. That is, a particular crystal may operate at a number of frequencies in the same way that a violin string can vibrate at a number of different frequencies called *overtones*. The crystal's vibration may range from a few thousand to many millions of cycles per second. Generally speaking, the smaller the crystal the higher the resonant frequencies at which it can vibrate. Crystals at the high-frequency end of the scale may be less than one millimeter thick. Thus we see that one of the limitations of crystal resonators is related to our ability to cut crystals precisely into very small bits.

The crystal resonator is incorporated into a feedback system that operates in a way similar to the one discussed on page 37. The system is self regulating, so the crystal output frequency is always at or near its resonant frequency. The first crystal clocks were enclosed in cabinets 3 meters high, 2½ meters wide, and 1 meter deep, to accommodate the various necessary components. Today quartz-crystal *wrist watches* are available commercially—which gives some indication of the great strides made in miniaturization of electronic circuitry over the past few years.

The best crystal clocks will keep time to one millisecond per month, whereas lower quality quartz clocks may drift a millisecond or so in several days. There are two main reasons that the resonant frequency of a quartz oscillator drifts. First, the frequency changes with temperature; and second, there is a slow, long-term drift that may be due to a number of things, such as contamination of the crystal with impurities, changes inside the crystal caused by its vibration, or other aspects of "aging."

Elaborate steps have been taken to overcome these difficulties by putting the crystal in a temperature-controlled "oven," and in a contamination-proof container. But just as in the case of Shortt's two-pendulum clock, a point of diminishing return arrives where one must work harder and harder to gain less and less.

VIOLIN STRING

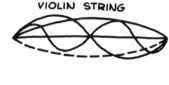

QUARTZ CRYSTAL CAPSULE

DIGITAL DISPLAY

ELECTRONIC TIMING CIRCUITRY

BATTERY

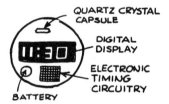

## ATOMIC CLOCKS—Q = 10⁵ — 10⁹

The next big step was the use of *atoms* (actually, at first, *molecules*) for resonators. One will appreciate the degree of perfection achieved with atomic resonators when he is told that these resonators achieve $Q$'s over *100 million.*

To understand this we must abandon Newton's laws, which describe swinging pendulums and vibrating materials, and turn instead to the laws that describe the motions of atoms and their interactions with the outside world. These laws go under the general heading of "quantum mechanics," and they were developed by different scientists, beginning about 1900. We shall pick up the story about 1913 with the young Danish physicist Niels Bohr, who had worked in England with Ernest T. Rutherford, one of the world's outstanding experimental physicists. Rutherford bombarded atoms with alpha particles from radioactive materials and came to the conclusion that the atom consists of a central core surrounded by orbiting electrons like planets circling around the sun.

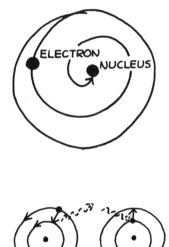

But there was a very puzzling thing about Rutherford's conception of the atom: Why didn't atoms eventually run down? After all, even the planets, as they circle the sun, gradually lose energy, moving in smaller and smaller circles until they fall into the sun. In the same manner the electron should gradually lose energy until it falls into the core of the atom. Instead, it appeared to circle the core with undiminished energy, like a perpetual motion machine, until suddenly it would jump to another inner orbit, releasing a fixed amount of energy. Bohr came to the then revolutionary idea that the electron did not *gradually* lose its energy, but lost energy in "lumps" by jumping between definite orbits, and that the energy was released in the form of radiation at a particular *frequency.*

Conversely, if the atom is placed in a radiation field it can *absorb* energy only in discrete lumps, which causes the electron to jump from an inner to an outer orbit. If there is no frequency in the radiation field that corresponds to the energy associated with an allowed jump, then no absorption of energy can take place. If there is such a frequency, then the atom can absorb energy from the radiation field.

EMIT          ABSORB

The frequency of the radiation is related to the lump or quantum of energy in a very specific way: The bigger the quantum of energy, the higher the emitted frequency. This energy-frequency relationship, combined with the fact that only certain quanta of energy are allowed—namely, the ones associated with electron jumps between specific orbits—is an important phenomenon for clockmakers. It suggests that we can use atoms as *resonators,* and furthermore that the emitted or resonant frequency is a property of the atom itself.

This is a big advance because now we don't have to be concerned with such things as building a pendulum to an exact length or cutting a crystal to the correct size. The atom is a natural, non-man-made resonator whose resonant frequency is practically immune to the temperature and frictional effects that plague mechanical clocks. The atom seems to be approaching the ideal resonator.

But we are still a long way from *producing* an atomic resonator. How are we to count the "ticks" or measure the frequency of such a resonator? What is the best atom to use? How do we get the electron in the chosen atom to jump between the desired orbits to produce the frequency we want?

We have partially answered these questions in the section on "Pushing $Q$ to the Limit," on page 37. There we described a feedback system consisting of three elements—an oscillator, a high-$Q$ resonator, and a feedback path. The oscillator produces a signal that is transmitted to the high-$Q$ resonator, causing it to vibrate. This vibration in turn, through suitable electronic circuitry, generates a signal proportional to the magnitude of the vibration that is fed back to the oscillator to adjust its frequency. This process goes around and around until the high-$Q$ resonator is vibrating with maximum amplitude; that is, it is vibrating at its resonant frequency.

In the atomic clocks that we shall be discussing, the oscillator is always a crystal oscillator of the type discussed in the previous section, whereas the high-$Q$ resonator is based upon some natural resonant frequency of different species of atoms.

In a sense, atomic clocks are the "offspring" of Shortt's two-pendulum clock, where the crystal oscillator corresponds to one pendulum and the high-$Q$ resonator to the other.

### The Ammonia Resonator—$Q = 10^5 - 10^6$

In 1949, the National Bureau of Standards announced the world's first time source linked to the natural frequency of atomic particles. The particle was the ammonia molecule, which has a natural frequency at about 23,870 MHz. This frequency is in the microwave part of the radio spectrum, where radar systems operate. During World War II, great strides had been made in the development of equipment operating in the microwave region, and attention had been focused on resonant frequencies such as that of the ammonia molecule. So it was natural that the first atomic frequency device followed along in this area.

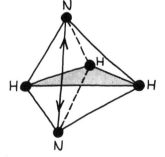

The ammonia molecule consists of three hydrogen atoms and one nitrogen atom in the shape of a pyramid, with the hydrogen atoms at the base and the nitrogen atom at the top. We have seen how the rules of quantum mechanics require that atoms emit and absorb energy in discrete quanta. According to these rules the nitrogen atom can jump down through the base of the pyramid and appear on the other side, thus making an upside-down pyramid. As we might expect, it can also jump back through the base to its original position. The molecule can also spin around different axes of rotation: The diagram shows one possibility. Each allowed rotation corresponds to a different energy state of the molecule. If we carefully inspect one of these states, we see that it actually consists of two distinct, but closely spaced, energy levels. This splitting is a consequence of the fact that the nitrogen atom can be either above or below the base of the pyramid. The energy difference between a pair of levels corresponds to a frequency of about 23,870 MHz.

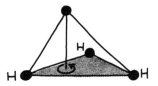

To harness this frequency a feedback system is employed consisting of two "pendulums": a quartz-crystal oscillator and the ammonia molecules. The quartz-crystal oscillator generates a frequency near that of the ammonia molecule. We can think of this signal as a weak radio signal being broadcast into a chamber of ammonia molecules. If the radio signal is precisely at the resonant frequency of the ammonia molecules, they will oscillate and strongly absorb the radio signal energy, so little of the signal passes through the chamber. At any other frequency the signal will pass through the ammonia, the amount of absorption being proportional to the difference between the radio signal frequency and the resonant frequency of the ammonia. The radio signal that gets through the ammonia is used to adjust the frequency of the quartz-crystal oscillator to that of the ammonia resonant frequency. Thus the ammonia molecules keep the quartz-crystal oscillator running at the desired frequency.

The quartz-crystal oscillator in turn controls some display device such as a wall clock. Of course, the wall clock runs at a much lower frequency—usually 60 Hz, like an ordinary electric kitchen clock. To produce this lower frequency the crystal frequency is reduced by electronic circuitry in a manner similar to using a train of gears to convert wheels running at one speed to run at another speed.

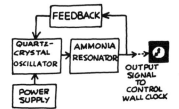

Although the resonance curve of the ammonia molecule is very narrow compared to previously used resonators, there are still problems. One is due to the collision of the ammonia molecules with one another and with the walls of the chamber. These collisions produce forces on the molecules that alter the resonant frequency.

Another difficulty is due to the motions of the molecules—motions that produce a "Doppler shift" of the frequency. We have observed Doppler frequency shifts when we listen to the whistle of a train as it approaches and passes us. As the train comes toward us, the whistle is high in pitch, and then as the train passes by, the pitch lowers. This same effect applies to the speeding ammonia molecules and distorts the results. Turning to the cesium atom instead of the ammonia molecule minimizes these effects.

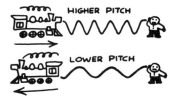

### The Cesium Resonator—$Q = 10^7 - 10^8$

The cesium atom has a natural vibration at 9,192,631,770 Hz, which is, like that of the ammonia molecule, in the microwave part of the radio spectrum. This natural vibration is a property of the atom itself, in contrast to the ammonia natural frequency, which results from the interactions of four atoms. Cesium is a silvery

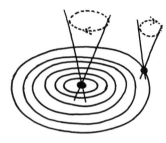

metal at room temperature. The core of the atom is surrounded by a swarm of electrons, but the outermost electron is in an orbit by itself. This electron spins on its axis, producing a magnetic field; we could thus think of the electron as being a miniature magnet. The core or nucleus of the cesium atom also spins, producing another miniature magnet, each magnet feeling the force of the other.

These two magnets are like spinning tops wobbling around in the same way the earth wobbles because of the pull of the moon. (This wobbling motion of the earth is discussed more fully on page 68.) If the two magnets are aligned with their "north" poles in the same direction, the cesium atom is in one energy state; and if they are aligned in opposite directions, the atom is in a different energy state. The difference between these two energy states corresponds to a frequency of 9,192,631,770 Hz. If we immerse the cesium atoms in a "bath" of radio signals at precisely this frequency, then the outside spinning electron can "flip over," either absorbing energy or emitting energy.

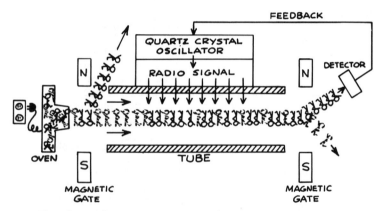

The figure illustrates the operation of the cesium-beam frequency standard. On the left is a small electric "oven" that heats the cesium atoms so that they are "boiled out" through a small opening into a long, evacuated tube. The atoms travel down the tube like marching soldiers, thus avoiding collisions with each other— which we recall was one of the difficulties with the ammonia resonator. As the atoms pass along the tube they come to a "gate," which is in reality a special magnetic field that separates the atoms into two streams according to whether their electron is spinning in the same direction as the nucleus or the opposite direction. Only one kind of atom is allowed to proceed down the tube, while the others are deflected away. The selected (gray) beam then passes through a section of the tube where the particles are exposed to a radio signal very near 9,192,631,770 Hz. If the radio signal is precisely at the resonant frequency, then large numbers of atoms will change their energy state, or "flip over."

The atoms then pass through another magnetic gate at the end of the tube. Those atoms that have changed energy state while passing through the radio signals are allowed to proceed to a detector at the end of the tube, while those that did not change state are deflected away from the detector. When the radio frequency is equal to the resonant frequency, the greatest number of atoms will reach the detector. The detector produces a signal that is related to the number of atoms reaching it, and this signal is fed back to control the radio frequency through a crystal oscillator so as to maximize the number of atoms reaching the detector—which, of course, means that the radio signal is at the cesium atom's resonant frequency. In this way the crystal oscillator frequency is tied to the resonant frequency of the cesium atom. The whole process, which is automatic, is much like carefully tuning in a radio so that the receiver gets the loudest and clearest signal; when this happens, the receiver is exactly "on-frequency" with the signal sent.

We have seen that one of the difficulties with the ammonia resonator is avoided by having the cesium atoms march down the tube with as little interaction as possible. The spread in frequency caused by the Doppler shift is minimized by transmitting the radio signal at right angles to the beam of the cesium atoms, as shown in the figure; the cesium atoms are never moving toward or away from the radio signal, but always across it.

### One Second in 370,000 Years

Carefully constructed cesium-beam-tube resonators maintained in laboratories have $Q$'s over 100 million, whereas smaller, portable units, about the size of a piece of luggage, have $Q$'s of about 10 million. In principle, laboratory oscillators keep time to about *one second in 370,000 years*—if we could build one that would last that long. A few microseconds per year is what is really important, however, and that's the same ratio.

What accounts for this high $Q$ of a cesium resonator? In our discussion of $Q$ we saw that the frequency spread of the resonance curve decreases as the "decay" time increases. In fact, the spread is just one over the decay time ( $\frac{1}{\text{decay time}}$ ). In the case of the cesium-beam tube, the decay time is simply the time it takes the cesium atoms to travel the length of the tube. Laboratory cesium-beam tubes may be as long as four meters, and the cesium atoms boiling out of the electric oven travel down the tube at about 100 meters per second; so the cesium atom is in the tube about 0.04 second. Since the frequency spread is one over the time the atom spends in the tube, we obtain a frequency spread of 1/0.04, which equals 25 Hz. But the $Q$ is the resonant frequency divided by the frequency spread, or 9,192,631,770 Hz/25 Hz, or about 400 million.

$$\text{TIME IN TUBE} = \frac{\text{LENGTH OF TUBE}}{\text{SPEED OF ATOMS}} = 0.04 \text{ sec.}$$

$$\text{FREQUENCY SPREAD} = \frac{1}{\text{TIME IN TUBE}} = \frac{1}{0.04} = 25 \text{ Hz}$$

### Atomic Definition of the Second

Because of the smoothness with which the cesium resonator "ticks," the definition of the second based on astronomical observation was abandoned in 1967, and the second was redefined as the duration of 9,192,631,770 vibrations of the cesium atom. On page 64, there is a much fuller discussion of the events that led to the atomic definition of the second. As we shall also see in that chapter, national laboratories responsible for the ultimate generation of time information do not literally have a large atomic clock with a face and hands like a wall clock, but rather the "clock" consists of a number of components, one of which is a set of atomic oscillators whose job it is to provide accuracy and stability for the entire clock system.

### The Rubidium Resonator—$Q = 10^7$

The rubidium resonator, although of lower quality than the cesium resonator, is nevertheless important because it is relatively inexpensive compared to cesium resonators, and because it is more than adequate for many of today's needs. The device is based on a particular resonant frequency of rubidium atoms contained as a gas at very low pressure in a specially constructed chamber.

Atoms, like crystals, have more than one resonant frequency. One of the rubidium resonant frequencies is excited by an intense beam of light, and another resonant frequency is excited by a radio wave in the microwave frequency region. As the light shines through the glass bulb containing rubidium gas, atoms in the "correct" energy state will absorb energy. (The situation is similar to the cesium atoms passing through the radio signal, where only those atoms with the outer electron spinning in the proper direction could absorb the radio signal and flip over to produce a different energy state.)

The microwave radio signal, when it is at the resonant frequency of the rubidium atom, converts the maximum number of atoms into the "correct" kind to absorb energy from the light beam. And as more of the atoms in the bulb are converted into the correct kind, they absorb more of the energy of the light beam; thus when the light beam is most heavily absorbed, the microwave signal is at the desired frequency. Again, as in the case of the cesium-beam tube, the amount of light that shines through the beam is detected and used to generate a signal that controls the microwave frequency to make the light beam reach minimum value.

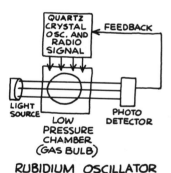

RUBIDIUM OSCILLATOR

Rubidium oscillators have $Q$'s of around 10 million, and they keep time to about one millisecond in a few months. But like crystal oscillators, they drift slowly with time and must occasionally be reset with reference to a cesium oscillator. This drift is due to such things as drift in the light source and absorption of rubidium in the walls of the storage bottle.

### The Hydrogen Maser—$Q = 10^9$

In the cases of the three atomic resonators we have discussed —ammonia, cesium, and rubidium resonators—we observe the resonant frequency indirectly. That is, we measure, in the case of the cesium oscillator, the number of atoms reaching the detector. In the cases of the ammonia and rubidium devices we measure the

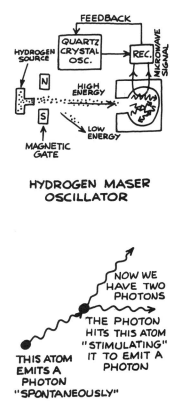

amount of signal absorbed as the signal passes through atoms and molecules. Why not observe the atomic radio or optical signal directly? The next device we shall discuss—the hydrogen maser— does just that.

The man who developed the maser, Dr. Charles H. Townes, an American scientist, was not working on an oscillator at all; rather he was seeking a way to amplify microwave radio signals. Hence the name *maser*, which although now a common noun in the dictionary, is simply an acronym for *M*icrowave *A*mplification by *S*timulated *E*mission of *R*adiation. But as we have seen, anything that oscillates or swings with a definite period or frequency can become the basis of a timekeeping device or clock. The resonator in the hydrogen masers is the hydrogen atom, which has, among others, a particular resonant frequency of 1,420,405,752 Hz.

In a manner similar to that of the cesium-beam tube, hydrogen gas drifts through a magnetic gate that allows only those atoms in an energy-emitting state to pass. Those atoms making it through the gate enter a quartz-glass storage bulb a few inches in diameter. The bulb is coated inside with a material similar to that used on nonstick cookware. For reasons not entirely understood, this coating reduces the frequency-perturbing effects caused by collisions of the hydrogen atoms with the wall of the bulb. The atoms stay in the bulb about one second before leaving; and thus their effective decay time is about one second, as compared to 0.04 second for the cesium-beam tube. This longer decay time results in a $Q$ about ten times higher than that of the cesium beam oscillator, even though the resonant frequency is lower.

If the bulb contains enough hydrogen atoms in the energy-emitting state, "self-oscillation" will occur in the bulb. According to the laws of quantum mechanics, an atom in an energy-emitting state will, eventually, *spontaneously* emit a packet of radiation energy. Although it is not possible to know in advance which particular atom will emit energy, if there are enough atoms in the quartz bulb eventually one of them spontaneously emits a packet of energy, or *photon*, at the resonant frequency. If this photon hits another atom in an energy-emitting state, that atom may be "stimulated" to release its energy as another photon of exactly the same energy—and therefore the same frequency—as the one that started the process. The remarkable thing is that the "stimulated" emission is in step with the radiation that produced it. The situation is similar to that of a choir in which all members are singing the same word at the same time, rather than the same word at different times.

We now have two photons bouncing around inside the bulb, and they will interact with other energy-emitting atoms; so the whole process escalates like a falling house of cards. Since all of the photons are in step, they constitute a microwave radio signal at a particular frequency, which is picked up by a receiver. This signal keeps a crystal oscillator in step with the resonant frequency of the energy-emitting hydrogen atoms. Energy is supplied by a constant stream of hydrogen atoms in their high-energy state, and thus, a continuous signal results.

Although the $Q$ of the hydrogen resonator is higher than that of the cesium-beam resonator, its accuracy is not as great today because of the unsolved problem of accurately evaluating and minimizing the frequency shift caused by the collisions between the hydrogen atoms and the wall of the quartz bulb.

## CAN WE ALWAYS BUILD A BETTER CLOCK?

We have seen that the $Q$ of the resonator is related to its decay time. For the atomic resonators we have discussed, the decay time was largely determined by the length of time the atom spends in some sort of container—a beam tube or a bulb. Historically, the trend has been toward resonators with higher and higher resonant frequencies. But this turns out to have an impact on the decay time. As we said, atoms in energy-emitting states can, given sufficient time, release spontaneously a burst of energy at a particular frequency. According to the rules of quantum mechanics, the decay time decreases rapidly with increasing frequency. At such high frequencies, the average time for spontaneous emission—or natural lifetime—may be considerably smaller than the time that the atom spends in the container.

In the case of the cesium-beam tube and hydrogen in the quartz bulb, the natural lifetimes of the atoms are considerably longer than the containment times in the bulb or beam tube; but at much higher frequencies this may not be the case. So it would appear that the recent trend toward basing resonators on higher and higher atomic resonant frequencies may eventually reach some upper limit. But that limit is not yet in sight. As we shall see in the final chapter, there are suggestions of even more distant horizons where clock resonators may be based on emissions from the nucleus of the atom itself.

For the present, the only limit to building better and better clocks would appear to be the upper reaches of man's ingenuity in coping with the problems that inevitably arise when a particular path is taken. So it is man's imagination, not nature, that dictates the possibilities for the foreseeable future.

The atomic resonators we have discussed are, of course, far too cumbersome and expensive for any but scientific, laboratory, and similar specific uses; and their operation and maintenance require considerable expertise. But a few years ago the same would have been said of the quartz-crystal oscillator, which has now become common in wrist watches. Although it seems unlikely at this time, who is to say that there may not be some breakthrough that will make some sort of atomic clock much more practicable and widely available than it is today?

**DECAY TIME DECREASES WITH INCREASING FREQUENCY**

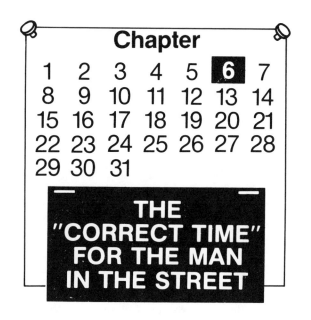

# Chapter

| 1 | 2 | 3 | 4 | 5 | **6** | 7 |
| 8 | 9 | 10 | 11 | 12 | 13 | 14 |
| 15 | 16 | 17 | 18 | 19 | 20 | 21 |
| 22 | 23 | 24 | 25 | 26 | 27 | 28 |
| 29 | 30 | 31 | | | | |

## THE "CORRECT TIME" FOR THE MAN IN THE STREET

Thus far we have concentrated on the technical developments that led to improved timekeeping, and we have seen how these developments were utilized in scientific and national standards laboratories, where the utmost accuracy and stability are required. Now we shall turn to the more pedestrian timepieces, which we can carry in our pockets or wear on our wrists. These watches operate in a manner similar to their laboratory cousins, but they are less accurate for reasons of economy, size, and convenience.

### THE FIRST WATCHES

The word *watch* is a derivation from the Anglo-Saxon *wacian* meaning "to watch," or "to wake." Probably it described the practice of the man keeping the "night watch," who carried a clock through the streets and announced the time, as well as important news—or simply called out, "Nine o'clock and all's well."

Early clocks were powered by weights suspended from a rope or chain—an impractical scheme for portable timepieces. The breakthrough came in about 1500, when Robert Henlein, a German locksmith, realized that a clock could be powered by a coiled brass or steel spring. The rest of the clock was essentially the "foliot" mechanism already discussed on page 26, which was very sensitive to whether it was upright in position or lying on its side.

In 1660 the English physicist Robert Hooke toyed with the idea that a straight metal spring could act as a resonator in a clock; and in 1675 the Dutch physicist and astronomer Christian Huygens employed this principle in the form of a metal spiral

ANCHOR
ESCAPEMENT

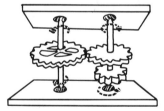

spring connected to a rotating balance wheel; energy flowed back and forth between the moving wheel and the coiled spring.

Hooke is also credited with development of a new kind of escapement, the "anchor" escapement—so called because of its shape—which with the help of its "escape wheel" delicately transferred energy to the resonator of the clock. With these developments, the accuracy of clocks improved to the point where the minute hand was added in the latter part of the 1600's.

The history of watches, up until about the middle of the 17th century, was essentially one of gradually improving the basic design of the first watches—most of which were so large that we would probably refer to them today as clocks. As Brearley explains in *Time Telling through the Ages*: "Back in 1650 it was some job to figure out the number of teeth in a train of wheels and pinions for a watch, to determine their correct diameters, to ascertain the number of beats of the escapement per hour, and then design a balance wheel and hair spring that would produce the requisite number; to determine the length, width, and thickness of a main spring that would furnish enough and not too much power to drive the mechanism; and finally, with the very crude and inadequate tools then available, to execute plans and produce a complete watch that would run and keep time—even approximately."

One significant development occurred in 1704, when Nicholas Facio, of Basel, Switzerland, introduced the jeweled bearing. Up to that time, the axles of the gears rotated in holes punched in brass plates—which considerably limited the life and accuracy of the watch.

Before the middle of the 17th century the production of clocks and watches was largely the work of skilled craftsmen, principally in England, Germany, and France, although it was the Swiss craftsmen who introduced nearly all of the basic improvements in the watch. The watchmaker—or "horological artist," as he was called—individually designed, produced, and assembled all parts of each watch, from the jeweled bearings and pinioned wheels to the face, hands, and case. In some cases an horologist might take an entire year to build a single timepiece.

In Switzerland, however, and later in the United States, watchmakers became interested in ideas that the industrial revolution was bringing to gunsmithing and the making of other mechanisms. The manufacture of identical and interchangeable parts that could be used in making and repairing watches made possible the mass production of both expensive and inexpensive watches. Turning to this kind of standardization, Switzerland rapidly became known throughout the world as the center for fine watchmaking. About 6000 watches were produced in Geneva in 1687, and by the end of the 18th century Geneva craftsmen were producing 50,000 watches a year. By 1828 Swiss watchmakers had begun to make watches with the aid of machinery, and mass production of watches at a price that the average man could afford was assured.

But it was in the United States that the idea of machine-produced interchangeable parts finally resulted in a really inexpensive watch that kept reasonably good time. After many false starts and efforts by various persons that met with little success, R. H. Ingersoll launched the famous "dollar watch" about the end of the 19th century. A tremendous success, it sold in the millions throughout the next quarter century or more. The first were pocket watches, encased in a nickel alloy; but as the wrist watch gained acceptance and popularity in the 1920's, Ingersoll also manufactured both men's and ladies' wrist watches.

## MODERN MECHANICAL WATCHES

It was style consciousness that was largely responsible for continued changes and improvements in the watch mechanism. The challenge of producing watches small and light enough to be pinned to the sheer fabrics of ladies' daytime and evening dresses without pulling the dress lines out of shape resulted in the dainty, decorative pendant watches popular in the early 1900's. Designing works that would fit into the slim, curved wrist-watch case that became increasingly popular with men was a major achievement after World War I.

## TIGER

©KING FEATURES SYNDICATE, INC. 1977 — REPRINTED BY PERMISSION OF KING FEATURES SYNDICATE

Accuracy, stability, and reliability also remained important goals. The multiplying railroad lines, with their crack trains often running only minutes apart by the latter half of the 19th century, helped to create a strong demand for accurate, reliable watches. Every "railroader," from the station manager and dispatcher to the engineer, conductor, and track repair crew with their motorcar, had to know the time, often to the part of a minute. A railroad employee took great pride in his watch—which he had to buy himself and which had to meet specified requirements.

Before electronic watches entered the scene, the Union Pacific Railroad required that all watches have 21 jewels and that they be a certain minimum size. Today electronic wrist watches are allowed, but whatever the type, each morning a railroader's watch must be checked against a time signal coming over a telegraph wire or the telephone, and it must be within 30 seconds of the correct time. As an additional safety measure, watches are checked on the job by watch inspectors, who appear unannounced. Time, to the railroads, is still a very serious matter.

Today's mechanical watch is a marvel of the art of manufacturing and assembling of tiny parts. The balance wheel in a ladies' wrist watch has a diameter about the same as that of a matchhead, and the escapement ticks over one hundred million times a year, while the rim of the balance wheel travels over 11,200 kilometers miles in its back-and-forth journey. The balance wheel is balanced and its rate adjusted by over a dozen tiny screws around its rim. Some 30,000 of these screws would fit into a ladies' thimble, and the jewels may be as small as specks of pepper. It's no wonder that the tiniest piece of dust can stop a watch or seriously impair its motion.

Even oiling a watch is a delicate operation. One single drop of oil from a hypodermic syringe is enough to lubricate over a thousand jeweled bearings. An amazing variety of substances have been used for lubrication, ranging from porpoise-jaw oil to today's modern synthetic oils.

"Every night, when he winds up his watch, the modern man adjusts a scientific instrument of a precision and delicacy unimaginable to the most cunning artificers of Alexandria in its prime."
—Lancelot Hogben

## ELECTRIC AND ELECTRONIC WATCHES

A very big step in the development of the watch occurred in 1957, with the introduction of the electric watch. This watch was essentially the same as its mechanical predecessor, except that it was powered by a tiny battery instead of a spring. Two years later, in 1959, a watch was introduced with the balance wheel replaced by a tiny tuning fork. Historically we have seen that the quality factor, or $Q$, of resonators increases with resonance frequency. The balance wheel in mechanical watches swings back and forth a few times a second, but the tuning fork vibrates several hundred times a second, with a $Q$ around 2,000—20 times better than the average balance wheel resonator. Such watches can keep time to a minute in a month. The tuning fork's vibrations are maintained by the interaction between a battery-driven, transistorized oscillating circuit and two tiny permanent magnets attached to the ends of the tuning fork.

## THE QUARTZ CRYSTAL WATCH

The quartz-crystal wrist watch, which is a miniature version of the quartz-crystal clock discussed on page 40, is the latest step in the evolution of watches. Its development was not possible until the invention of the integrated circuit—the equivalent of many hundreds of thousands of transistors and resistors in an area only a centimeter or less on a side. These circuits can carry out the many complex functions of a watch, one of the most important being the electronic counting of the vibrations of the quartz-crystal resonator.

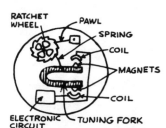

RATCHET WHEEL — PAWL — SPRING — COIL — MAGNETS — COIL — ELECTRONIC CIRCUIT — TUNING FORK

THE VIBRATING FORK PUSHES A SMALL SPRING AGAINST THE RATCHET WHEEL WHICH MOVES THE HANDS OF THE WATCH. THE PAWL KEEPS THE RATCHET WHEEL FROM MOVING BACKWARDS. THE TWO COILS, CONTROLLED BY AN ELECTRONIC CIRCUIT, INTERACT WITH THE TWO MAGNETS TO KEEP THE TUNING FORK VIBRATING.

The first quartz wrist watches utilized the "hands" type of display, adapted from existing watches. But later versions became available with no moving parts at all. The hands have been replaced by digital time "readouts" in hours, minutes, and seconds formed from small luminous elements that are entirely controlled by electrical signals. Quartz watches today are accurate to one minute a year, but they are in a very early developmental stage, and it is too early to predict what ultimately may be achieved.

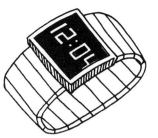

## HOW MUCH DOES "THE TIME" COST?

Clocks and watches are big business. In 1974, 200 million clocks and watches were sold worldwide for a price of four billion dollars, but the basic value of every one of these timepieces depends on its being "set" to the "correct" time, and then having access to a source against which it can be checked occasionally. So how much does "the time" cost?

For about 99 percent of the people who want to know what time it is—or to clock the duration of time—a clock or watch that "keeps time" within a minute or so a day is acceptable. The familiar and inexpensive wall or desk clock driven by the electric current supplied by the power company is completely adequate for the vast majority of people; few persons recognize a need for a more "refined" time. Using only his eyes and fingers, a human being has not the manual dexterity to set a clock or watch to an accuracy of better than a second or so, even if he has the time and patience to do it.

"Losing" the time altogether, when a clock or watch stops, is no problem to most people. One simply dials the telephone company time service or consults another of the many possible suppliers of the "correct" time. In short, for nearly everyone, in nearly all circumstances, the wide choice of clocks and watches available in the local drug or department store at prices of $5.00 and up is sufficient to meet everyday needs.

But let's suppose that a man is going into a remote area on a trip, where he has no radio receiver and will have no contact with other people for three or four weeks. If it's a fishing trip, he probably doesn't care whether his watch loses or gains a few minutes a day. He'll still probably make connections with the pilot who is flying in to pick him up at the end of the period.

But let's suppose that the man is to make certain observations at certain times of day, and that a scientific laboratory is depending on the information he gathers, and it's important to the laboratory that the time the information is recorded be correct within a tolerance of one minute. Or perhaps the man has a radio transmitter and is only one of several men in the field, each of whom is to send in a report at certain times each day. Then he will need a more accurate and dependable—and more expensive—watch. A $300 to $400 watch that's waterproof and shockproof, and that has proved to keep time without resetting—losing, say, no more than 30 seconds in six months or so—should serve him nicely. Especially if he has a radio receiver with which he can pick up a time broadcast occasionally, so he could check the time once in awhile and reset his watch if he needs to.

But then there is a surprising array of very common time users for whom any kind of watch or clock that can be read by the human eye or ear is as useless as a meterstick is to a lens grinder. Communications and power company engineers, scientific laboratory technicians, and many other special users of time and frequency information read this information with the help of an *oscilloscope* hooked up to sophisticated receiving instruments. Their clock may be driven by a quartz-crystal oscillator, which, although accurate to one millisecond per month, must be checked where more accurate time is required—often several times a day—by an even better oscillator. A quartz-crystal oscillator may cost as much as $2500, depending on its quality. It must have a special housing with controlled temperature and humidity, and it requires someone with special training in its care and use to look after and regulate it. Often a team of technicians read it, chart its performance, and adjust it as needed, every day.

These individuals, obviously, must have an even better time source than their quartz-crystal clocks in order to keep them telling the time accurately. This will be an atomic clock of some kind. Perhaps a rubidium frequency standard, that costs about $7500, or a cesium standard with a price tag of around $15,000. When a portable cesium standard is hand carried from its "home" to be checked and adjusted against another, similar standard—or against *the* NBS atomic frequency standard at the United States National Bureau of Standards or the official standard in another nation—it travels, usually by airplane, attended by a team of technicians who see that it is plugged into an electric circuit whenever possible, and that its batteries are kept charged for use when this is not possible.

A portable cesium standard weighs about 90 kilograms, and occupies about ⅓ of a cubic meter. Characteristically, it will not lose or gain one second in 3000 years. Such atomic standards are

CRYSTAL CLOCK
$2500

RUBIDIUM CLOCK
$7500

CESIUM CLOCK
$15,000

found in scientific laboratories, electronics factories, and even a few TV stations.

The primary frequency standard, at the laboratories of the National Bureau of Standards, in Boulder, Colorado, is much larger than the portable standards. Housed in its own special room, it is about 6 meters long and weighs over 3000 kilograms. The present model was completed in 1976 at a cost of about $300,000. It is used in conjunction with several smaller atomic "clocks" that monitor each other constantly and are the basis of the NBS time and frequency services. It is accurate to 1 second in 370,000 years.

So who needs the $300,000 clock? We all do. We need it to set our $15 watch. Everyone who uses a television set, a telephone, electric shaver, record player, vacuum cleaner, or clock depends ultimately on the precise time and timing information supplied by this $300,000 clock. Not to mention everyone whose daily activities are more or less regulated by and dependent on the working of hundreds of computers plugged into each other all across the nation—everything from airplane and hotel reservations to stock market quotations and national crime information systems.

"The time" is very inexpensive and easy to come by for many millions of average users, simply because relatively *few* users must have very expensive and much more refined, precise time. The remarkable accuracy and dependability of the common electric wall clock can be bought very cheaply only because very much more expensive clocks make it possible for the power company to deliver electricity at a very constant 60 cycles per second, or 60 hertz, day in and day out. The "time" as most of us know it is simply inexpensive crumbs from the tables of the few rich "gourmet" consumers of time and frequency information.

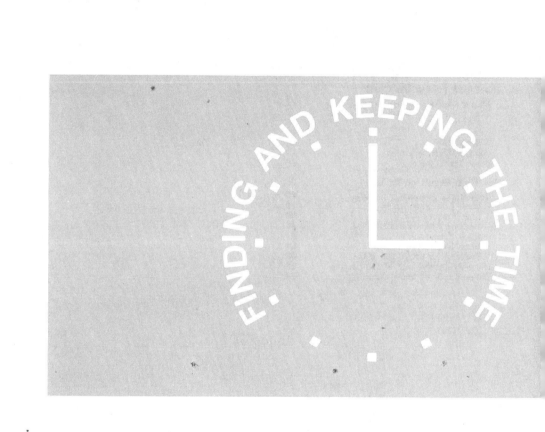

FINDING AND KEEPING THE TIME

## III
## FINDING AND KEEPING THE TIME

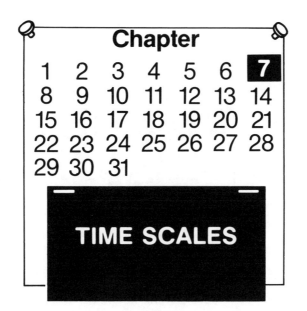

# Chapter 7

| | | | | | | |
|---|---|---|---|---|---|---|
| 1 | 2 | 3 | 4 | 5 | 6 | **7** |
| 8 | 9 | 10 | 11 | 12 | 13 | 14 |
| 15 | 16 | 17 | 18 | 19 | 20 | 21 |
| 22 | 23 | 24 | 25 | 26 | 27 | 28 |
| 29 | 30 | 31 | | | | |

## TIME SCALES

Length scales may measure inches or centimeters, miles or kilometers. Weight scales may be read in ounces or grams. When we speak of an ounce, we sometimes have to specify whether we mean avoirdupois weight or apothecaries' weight, for each is measured by a different *scale*. Nautical miles are not measured by the same scale as statute miles. Time, too, is measured by different scales for different purposes and by different users, and the scales themselves have been modified throughout history to meet changing needs or to gain greater accuracy.

### THE CALENDAR

The year, the month, and the day are natural units of time derived from three different astronomical cycles:

- The year—solar year—is the period of one complete revolution of the earth about the sun.
- The month is the time between two successive new moons.
- The day is the time between two successive "high" noons.

As man became more sophisticated in his astronomical measurements, he noticed that there were not an even number of days and months in the year. Early farmers in the Tigris-Euphrates Valley had devised a calendar with 12 months per year, each month being the average time between two new moons, or 29½ days. This adds up to 354 days per year, 11 days short of the year we know.

Before long the farmers noticed that their planting times were getting out of step with the seasons. To bring the calendar into conformity with the seasons, extra days and months were added, at first on a rather irregular basis, and later at regular intervals over a 19-year cycle.

The Egyptians were the first to recognize that the solar year was close to 365 days, and that even this calculation needed adjustment by adding one extra day every four years. However, the Egyptian astronomers could not persuade the rulers to add the extra day every fourth year, so the seasons and the calendar slowly drifted out of phase. It was not until some two centuries later that Julius Caesar, in 46 B.C., instituted the 365-day year adjusted for leap years. But even this adjustment isn't quite correct; a leap year every four years amounts to an over-correction on the average of 12 minutes every solar year. Some thousand years after Julius Caesar established his calendar, this small yearly error had accumulated to about six days; and important religious holidays such as Easter were moving earlier and earlier into the season.

By 1582, the error had become so great that Pope Gregory XIII modified the calendar and the rules for generating it. First, years initiating a new century, not divisible by 400, would not be leap years. For example, the year 2000 will be a leap year because it is divisible by 400, but the year 1900 was not. This change reduces the error to about one day in 3,300 years. Second, to bring the calendar back into step with the seasons, October 4 of 1582 was followed by October 15, removing 10 days from the year 1582.

With the adoption of the Gregorian Calendar, the problem of keeping the calendar in step with the seasons was pretty well solved. But we still have the awkward fact that the numbers of days and months in the year are not commensurate with the period of the earth's rotation around the sun. Thus, as long as we base our calendar upon these three astronomical cycles we will always be stuck with the kind of situation we have now, with different numbers of days in the months and the years.

## The Solar Day

We have just seen that because there is not an integral number of days or months in the year we have a rather ragged calendar. But there are more problems. As man improved his ability to measure time, he noticed that the time of day as measured by a sundial could vary from the "norm" as much as 15 minutes in February and November. There are two primary reasons for this:

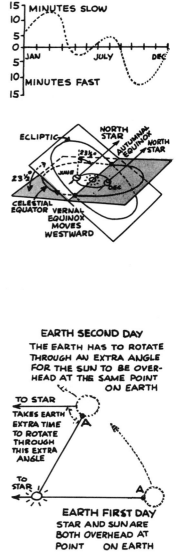

- The earth does not travel around the sun in a circle, but in an ellipse. When the earth is nearer the sun—in winter, in the northern hemisphere—it travels faster in orbit than when it is farther away from the sun—in summer.
- The axis of the earth's rotation is tilted at an angle of about 23½° with respect to the plane which contains the earth's orbit around the sun.

Together these two facts account for the discrepancies in February and November. Because of this variation, a new day called the "mean solar day" was defined. The mean solar day is simply the average length of all of the individual solar days throughout the year. The sketch shows how the length of the solar day leads and lags behind the mean solar day throughout the year.

## The Stellar or Sidereal Day

We have defined the solar day as the time between two "high" noons, or upper transits of the sun. But what if we measured the time between two upper transits of a star? Does the "star" day equal the solar day? No. We would find that the star would appear at upper transit a little earlier the second night. Why? Because the earth, during the time it is making one rotation about its axis, has moved some distance, also, in its journey around the sun. The net effect is that the mean solar day is about four minutes longer than the day determined by the star. The day determined by the star is called the *sidereal day*.

Unlike the solar day, the sidereal day does not vary in length from one time of the year to another; it is always about four minutes shorter than the *mean* solar day, regardless of the time of year or season.

Why is its length so much more constant? Because the stars are so far away from the earth that the tilt of the earth's axis and the elliptical orbit of the earth around the sun can be ignored. To put it differently, if we were looking at the earth from some distant star, we would hardly be able to discern that a tilted earth was moving around the sun in an elliptical orbit. In fact, the mean solar day itself can be more easily measured by observing the stars than it can by observing the sun.

## Earth Rotation

There still remains one final area of uncertainty in the astronomical time scale. Does the earth itself rotate uniformly? There were suspicions as early as the late 17th century that it does not. The first Astronomer Royal of England, John Flamstead, suggested in 1675 that since the earth is surrounded by water and air, whose distribution across the surface of the earth changes with time, its rotation rate might change from season to season.

A more definitive clue was obtained by the British astronomer Edmund Halley, after whom the famous comet was named. In 1695 Halley noticed that the moon was ahead of where it should have been. Either the earth was slowing down in its rotational rate or the moon's orbit had not been properly predicted. The moon's orbit was carefully recalculated, but no error was found.

The evidence continued to mount. Near the beginning of the 20th century, Simon Newcomb, an American astronomer, concluded that during the past two centuries the moon had been at times ahead of, and at times behind, its predicted position. By 1939 it seemed clear that the earth's rotation was not uniform. Not only was the moon not appearing where it was supposed to be, but the planets, too, were not in their predicted places. The obvious explanation was that the earth's rotation was not uniform.

B.C. BY PERMISSION OF JOHNNY HART AND FIELD ENTERPRISES, INC.

With the development of atomic timekeeping in the early 1950's, it was possible to study the irregularities in earth rotation more carefully; for time obtained from atomic clocks is more uniform than earth time. These studies, along with observations such as those just mentioned, indicate that there are three main types of irregularities:

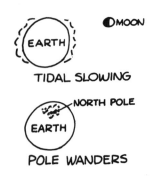

TIDAL SLOWING

POLE WANDERS

- *The earth is gradually slowing down;* the length of the day is about 16 milliseconds longer now than it was 1000 years ago. This slowing is due largely to frictional tidal effects of the moon on the earth's oceans. Indirect evidence from the annual growth bands on fossil corals suggests that the earth day was about 21 hours, six hundred million years ago.
- *The positions of the North and South poles wander around by a few meters from one year to the next.* Precise measurements

show that this wandering may produce a discrepancy as large as 30 milliseconds. This polar effect may be due to seasonal effects and rearrangements in the structure of the earth itself.

- *Regular and irregular fluctuations are superimposed on the slow decrease in rotation rate.* The regular fluctuations amount to a few milliseconds per year. In the spring the earth slows down, and in the fall it speeds up, because of seasonal variations on the surface of the earth, as first suspected by John Flamstead.

One possible explanation of this variation can be understood by recalling the figure of a skater spinning on one toe. As the skater draws his outstretched arms in toward his body, he spins faster. When he extends them, he slows down. This is so because rotational momentum cannot change unless there is some force to produce a change. The skater is an isolated spinning body with only a slight frictional drag caused by the air and the point of contact between the ice and the skate. When he pulls in his arms, his speed increases, so that his rotational momentum remains unchanged, and vice versa.

WINTER SLOWS DOWN

SUMMER SPEEDS UP

The earth is also an isolated spinning body. During the winter in the northern hemisphere, water evaporates from the ocean and accumulates as ice and snow on the high mountains. This movement of water from the oceans to the mountain tops is similar to the skater's extending his arms. So the earth slows down in winter; in the spring the snow melts and runs back to the seas, and the earth speeds up again.

One might wonder why this effect in the northern hemisphere is not exactly compensated by the opposite effect in the southern hemisphere during its change of seasons. The answer is that the land mass north of the equator is considerably greater than the mass south of the equator; and although there are compensating effects between the two hemispheres, the northern hemisphere dominates.

All of these effects that conspire to make the earth a somewhat irregular clock have led to the development of three different scales of time that are called *Universal Time*: UT0, UT1, and UT2.

- UT0 is the scale generated by the mean solar day. Thus UT0 corrects for the tilted earth moving around the sun in an elliptical orbit.
- UT1 is UT0 corrected for the polar motion of the earth.
- UT2 is UT1 corrected for the regular slowing down and speeding up of the earth in **winter** and **summer**. Each step from UT0 to UT2 produces a more uniform time scale.

## THE CONTINUING SEARCH FOR MORE UNIFORM TIME: EPHEM-
## ERIS TIME

As we have seen, time based upon the earth's rotation about its axis is irregular. We have also seen that because of this irregular rotation, the predicted times of certain astronomical phenomena such as the orbits of the moon and the planets are not always in agreement with the observations. Unless we assume that the moon and all of the planets are acting in an unpredictable, but similar, fashion, we must accept the only alternative assumption —that the earth's rotation is not steady.

Since this assumption seems the more reasonable—and has indeed been substantiated by other observations—we should assume that the astronomical events occur at the "correct" time, and that we should tie our time scale to these events rather than to earth rotation. This was in fact done in 1956, and the time based on the occurrence of these astronomical events is called *Ephemeris Time*.

## THE WIZARD OF ID

THE WIZARD OF ID BY PERMISSION OF
JOHNNY HART AND FIELD ENTERPRISES, INC.

## HOW LONG IS A SECOND?

The adoption of Ephemeris Time had an impact on the definition of the second, which is the basic unit for measuring time. Prior to 1956, the second was 1/86,400 of the mean solar day, since there are 86,400 seconds in a day. But we know that the second based on solar time is variable; so after 1956 and until 1967 the definition of the second was based upon Ephemeris Time. As a practical matter it was decided that the Ephemeris second should closely approximate the mean solar second, and so the Ephemeris second was defined as very near the mean solar second for the "tropical" year 1900. (Tropical year is the technical name for our ordinary concept of the year; it is discussed more fully on page 67.) Thus two clocks, one keeping Ephemeris Time (ET) and the other Universal Time (UT), would have been in close agreement in 1900. But because of the slowdown of the earth's rotation, UT was about 30 seconds behind ET by the middle of the century.

Ephemeris Time has the advantage of being uniform, and as far as we know it coincides with the uniform time that Newton had in mind when he formulated his laws of motion. The big disadvantage of Ephemeris Time is that it is not readily accessible because, by its very definition, we must wait for predicted astronomical events to occur in order to make a comparison. In other words, to obtain the kinds of accuracies that are required in the modern world, we must spread our astronomical observations over several years. For example, to obtain ET to an accuracy of 0.05 second requires making observations over a period of nine years!

UT seconds, by contrast, can be determined to an accuracy of a few milliseconds in one day because UT is based upon *daily* observations of the stars. But the fact remains that the UT second is a variable because of the irregularities in the earth's rotation rate. What was needed was a second that could be obtained accurately in a short time.

### "Rubber" Seconds

Scientists had developed workable atomic clocks by the early 1950's with accuracies never before realized. The problem was that even with the refinements and corrections that had been made in UT (Earth Time), UT and Atomic Time will get out of step because of the irregular rotation of the earth. The need persisted for a time scale that has the smoothness of Atomic Time but that will stay in approximate step with UT.

Such a compromise scale was generated in 1958. The de facto definition of the second was based on atomic time, but the time scale itself, called Coordinated Universal Time (UTC) was to stay in approximate step with UT2. It was further decided that there would be the same number of seconds in each year.

But this is clearly impossible unless the length of the second is changed periodically to reflect variations in the earth's rotation rate. This change was provided for, and the "rubber" second came into being. Each year, beginning in 1958, the length of the second, relative to the atomic second, was altered slightly, with the hope that the upcoming year would contain the same number of seconds as the one just passed. But as we have previously observed, the rotation rate of the earth is not entirely predictable; so there is no way to be certain in advance that the rubber second selected for a given year will be right for the year or years that follow.

In anticipation of this possibility it was further agreed that whenever UTC and UT2 differed by more than $\frac{1}{10}$ second, the UTC clock would be adjusted by $\frac{1}{10}$ second to stay within the specified tolerance.

But after a few years many people began to realize that the rubber-second system was a nuisance. Each year clocks all over the world had to be adjusted to run at a different rate. The problems were similar to those we might expect if each year the length of the centimeter was changed slightly, and all rulers—which were made of rubber, of course—had to be stretched or shrunk to fit the "centimeter of the year." Not only was it a nuisance to adjust the clocks, but in cases where high-quality clocks had to be adjusted, it was a very expensive operation. The rubber second was abandoned in favor of the atomic second.

### Atomic Time and the Atomic Second

The development of atomic frequency standards (see page 41) set the stage for a new second that could be determined accurately in a short time. In 1967 the second was defined in terms of the frequency of radiation emitted by a cesium atom. Specifically, by international agreement, the standard second was defined as the elapsed time of 9,192,631,770 oscillations of the "undisturbed" cesium atom. Electronic devices associated with an atomic clock count these oscillations and display the accumulating counts in the way that another clock counts the swings of a pendulum.

Now the length of the second could be determined accurately, in less than a minute, to a few billionths of a second. Of course, this new definition of the second is entirely independent of any earth motion; and so we are back to the same old, now-familiar problem: Because of the irregularity of the earth's rotation, Atomic Time and Earth Time (UT) will get out of step.

### The New UTC System and the Leap Second

To solve the problem of Atomic Time and Earth Time getting out of step, the "leap second" was invented in 1972. The leap second is similar to the leap year, when an extra day is added every fourth year to the end of February to keep the number of days in the year in step with the movement of the earth around the sun. Occasionally an extra second, the leap second, is added— or possibly subtracted—as required by the irregular rotation rate of the earth. More precisely, the rule is that UTC will always be within 0.9 seconds of UT1. The leap second is normally added to or subtracted from the last minute of the year, in December, or the last minute of June; and timekeepers throughout the world are notified by the Bureau International de l'Heure (BIH), in Paris, France, that the change is to be made. The minute during which the adjustment is made is either 59 or 61 seconds long.

---

BEFORE 1956

ONE SECOND =
MEAN SOLAR DAY
86,400
CALLED THE MEAN
SOLAR SECOND

1956-1967

ONE SECOND =
TROPICAL YEAR FOR 1900
31,556,925.9747
CALLED THE EPHEMERIS
SECOND

1967-

ONE SECOND =
9,192,631,770 OSCILLATIONS
OF THE "UNDISTURBED"
CESIUM ATOM
CALLED THE ATOMIC SECOND

In 1972—a leap year—two leap seconds were added, making it the "longest" year in modern times. Only one leap second was added in 1973, 1974, 1975, and 1976.

## THE LENGTH OF THE YEAR

Up to this point we have defined the year as the time it takes for the earth to make one complete journey around the sun. But actually there are two kinds of year. The first is the *sidereal year*, which is the time it takes the earth to circle around the sun with reference to the stars, in the same sense that the sidereal day is the time required for one complete revolution of the earth around its axis with respect to the stars. We can visualize the sidereal year as the time it would take the earth to move from some point, around its orbit, and back to the starting point—if we were watching this motion from a distant star. The length of the sidereal year is about 365.2564 mean solar days . . . Solar days (See page 61).

The other kind of year is the one we are used to in everyday life—the one that is broken up into the four seasons. This year is technically known as the *tropical year*, and its duration is about 365.2422 mean solar days, or about 20 minutes shorter than the sidereal year. The reason the two years are different lengths is that the reference point in space for the tropical year moves slowly itself, relative to the stars. The reference point for the tropical year is the point in space called the *vernal equinox*, which moves slowly westward through the background of stars. The sketch on page 61 shows how the vernal equinox is marked.

The celestial equator is contained in the plane that passes through the earth's equator, whereas the "ecliptic" is in the plane that passes through the earth's orbit around the sun. The vernal equinox and the autumnal equinox are the two points in space where the ecliptic and celestial equators intersect. The angle between the ecliptic and the celestial equator is determined by the tilt of the earth's axis of revolution to the plane of the ecliptic.

But why does the vernal equinox—and also the autumnal equinox—move slowly in space? For the same reason that a spinning top wobbles as it spins. The top wobbles because the earth's gravitation is trying to pull it on its side, while the spinning motion

SIDEREAL YEAR = 365.2564 MEAN SOLAR DAYS

TROPICAL YEAR = 365.2422 MEAN SOLAR DAYS

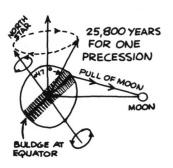

NORTH STAR

25,800 YEARS
FOR ONE
PRECESSION

23½°

PULL OF MOON

MOON

BULDGE AT
EQUATOR

produces a force that attempts to keep the top upright. Together the two forces cause the top to wobble, or "precess."

In the case of the earth, the earth is primarily the spinning top and the forces trying to topple it are the pulls of the moon and the sun; the moon produces the dominant force. If the earth were a perfect sphere with uniform density, there would be no such effect by the moon because all of the forces could be thought of as acting at the center of the earth. But because the earth spins, it bulges at the equator; and so there is an uneven distribution of mass, which allows the moon's gravitational field to get a "handle," so to speak, on the earth. The time for one complete precession is about 25,800 years, which amounts to less than one minute of arc per year. (One degree equals 60 minutes of arc.) But it is this slight yearly motion of the vernal equinox that accounts for the tropical year being about 20 minutes shorter than the sidereal year.

## THE KEEPERS OF TIME

Whatever the time scale and its individual advantages or idiosyncrasies, it is, of itself, simply a meter-stick, a basis for measurement. Before it's of any value, someone must put it to use; and someone must maintain and tend the instruments involved in the measurements. For as we've noted before, time is unique among the physical properties in that it is forever changing, and the meterstick that measures it can never be laid aside or forgotten about, to be activated only when someone wishes to use it.

B.C. BY PERMISSION OF JOHNNY HART AND FIELD ENTERPRISES, INC.

One can measure length or mass or temperature of an isolated entity, without consideration of continuity or needing to account for all of the space between two isolated entities. But every instant of time, in a sense, must be accounted for. If a month, a year, or a century doesn't "come out right" with respect to astronomical movements, it won't do simply to stop the clocks for the needed period of time—or move them ahead a certain amount—and start over. Every single second has its name on it, and each one must be accounted for, day after day, year after year, century after century. There has to be a general agreement among people and among nations on which time scale is to be used, and when alterations are to be made in "the time." This is a much greater and more elaborate undertaking than most persons realize.

The keepers of time make an important contribution to society, and even in days past they were held in high esteem. Ancient legends of primitive peoples often portray the great position of honor and trust occupied by the tender of the clock—also the tedious and sometimes irksome sense of responsibility felt by the tender, and the ignominy and condemnation heaped upon him by his fellow tribesmen when he failed in his duties. "The more things change, the more they are the same." The clock tender today is in much the same position, for although what passes for the "correct time" is all around us, all this information is of little value to the man responsible for maintaining the accuracy of "the clock"—or frequency standard. He must realize that his clock is an individual, unique in all the world, and that many operations involving time, money, and other people depend upon how well he tends his clock.

Whether the clock in question is one that regulates the activities of a radio or television station, tells the power company when it's putting out electricity at exactly 60 Hz, or providing location information on a ship at sea, the clock's "keeper" depends on a broader authority for *his* information. These authorities are both national and international.

### U.S. Timekeepers

There are two organizations in the United States primarily responsible for providing time and frequency information—the National Bureau of Standards (NBS) and the United States Naval Observatory (USNO), both organizations within the U. S. Government.

As we have seen, the present UTC time scale has both an astronomical and an atomic component: The length of the second is determined by atomic observations, whereas the number of seconds in the year is determined by astronomical observations. The atomic component can be divided into two parts—one part related to accuracy and another part related to stability.

Very roughly speaking, the USNO is responsible for the U. S. contribution to the *astronomical* part of UTC, and NBS is responsible for the U. S. contribution to the *accuracy* part of the *atomic* component, or length of the second; and both organizations provide input related to the *stability* part of the atomic component. The Bureau International de l'Heure, (BIH) in Paris, accumulates this data from many laboratories and observatories all over the world and calculates "the time."

The NBS input is generated by a system of atomic clocks in its laboratories in Boulder, Colorado. The system consists of a primary frequency standard, which is used to check the accuracy of a

TIME SCALES

1. ASTRONOMICAL COMPONENT
2. ATOMIC COMPONENT

   a. ACCURACY
   b. STABILITY

number of smaller, commercially-built secondary standards. The secondary standards run continuously and serve as a "flywheel" for the system.

To carry out its responsibilities, the USNO makes observations at its main facility in Washington, D. C., as well as in Richmond, Florida. The observations are made using a special telescope designed to measure the time when a given star passes overhead. By measuring the time between successive overhead passages of the star, the earth's rotation can be monitored and thus UT can be derived. To obtain the most precise measurements, a "photographic zenith tube" is used. With this device, the star is photographed automatically on a photographic plate that may contain the images of several stars, over one night's observation.

### The Bureau International de l'Heure

In an attempt to create a world time scale, some 70 nations of the world contribute data to the Bureau International de l'Heure, (BIH) in Paris, France. The BIH is the international headquarters for keeping time; and its responsibility is to take the information provided by the contributing nations to construct an international atomic time scale, the TAI scale. Some nations provide only astronomical information, others only stability information; and others provide accuracy, stability, and astronomical information. The time as determined by the BIH is just an average of all the various nations' time. It is also the responsibility of the BIH to determine when a leap second must be introduced.

From time to time NBS and other national timekeeping authorities make very exact comparisons of their clocks with the BIH clock. And with this clock as an agreed-upon standard, it is theoretically possible to keep all clocks in the world synchronized. But trying to realize this possibility is a constant challenge.

# Chapter

| 1 | 2 | 3 | 4 | 5 | 6 | 7 |
|---|---|---|---|---|---|---|
| **8** | 9 | 10 | 11 | 12 | 13 | 14 |
| 15 | 16 | 17 | 18 | 19 | 20 | 21 |
| 22 | 23 | 24 | 25 | 26 | 27 | 28 |
| 29 | 30 | 31 | | | | |

## THE CLOCK BEHIND THE CLOCK

A great many people carry "the time" around with them, in the form of a wrist watch. But what if the watch stops? Or what if two wrist watches show different time? How do the wearers know which is right—or whether either one is correct?

**TIGER**

Of course, they may ask a friend with a third watch—whose timepiece may or may not agree with one of the first two. Or one may dial the telephone company time service, or perhaps set his watch when he hears the time announcement on radio or television. The "correct time" seems to be all around us—on the wall clock at the drug store or court house, the outdoor time-and-temperature display at the local bank or shopping center. But a bit of observation will show that these sources are not always in agreement, even within a minute or so of each other, not to mention seconds or a fraction of a second. Which one is right? And where do *these* sources get the time? How do radio and television station managers know what time it is?

The answer is that there are, throughout the world, special radio broadcasts of accurate time information. Most of these broadcasts are at frequencies outside the range of ordinary United States AM radio; so one needs a special radio receiver to tune in the information. Many of these shortwave receivers are owned by radio and television stations, as well as by scientific laboratories in industry and government, and even by private citizens, such as boat owners, who need precise time information to navigate by the stars.

Of course, we come to the ultimate question: Where do these special broadcast stations go to find the time? And the answer is that many nations maintain the time by using very accurate atomic clocks combined with astronomical observations, as we discussed more fully in the previous chapter. All the time information from these various countries is constantly compared and combined, to provide a kind of "average world time," UTC, which is then broadcast by the special time and frequency radio stations located in various parts of the world.

## FLYING CLOCKS

Keeping the world's clocks synchronized, or running together, is an unceasing challenge. One of the most obvious ways to do it is simply to carry a third atomic clock between the master clock and the users' clocks. The accuracy of the synchronization depends primarily on the quality of the clock carried between the two locations and the time it takes to transport it. Usually the clock travels by airplane, carefully tended at all times by a team of technicians. Typically, the best quality portable atomic cesium clocks might drift between 0.1 and 1.0 microsecond per day. Carrying these portable atomic clocks is one of the main methods for comparing the time and frequency standards of the various nations with the BIH.

## TIME ON A RADIO BEAM

As early as 1840 it occurred to the English inventor Alexander Bain that it would be possible to send time signals over a wire. Bain obtained several patents, but it was not until a decade or so later that any serious progress was made in this direction. But before the middle of the 19th century the railroads were spreading everywhere, and their need for better time information and dissemination was critical. As the telegraph system developed by Samuel F. B. Morse grew with the railroads, systems were developed to relay time signals by telegraph, which automatically set clocks in all major railroad depots.

During the early part of the 20th century, with the development of radio, broadcasts of time information were initiated. In 1904 the United States Naval Observatory (USNO) experimentally broadcast time from Boston; and by 1910 time signals were being broadcast from an antenna located on the Eiffel Tower, in Paris. In 1912, at an international meeting held in Paris, uniform standards for broadcasting time signals were discussed.

In March of 1923, the National Bureau of Standards (NBS) began broadcasting its own time signal. At first there were only standard radio frequencies, transmitted on a regularly announced schedule from shortwave station WWV, located originally in Washington, D. C. One of the main uses of this signal was to allow radio stations to keep on their assigned frequencies, a difficult task during the early days of radio. In fact, one night in the 1920's the dirigible *Shenandoah* became lost in a winter storm over the eastern seaboard, and it was necessary for the New York radio stations to suspend transmissions so that the airship's radio message could be detected.

WWV was later moved outside Washington, D. C., to Beltsville, Maryland; and in 1966 to its present home at Fort Collins, Colorado, about 80 kilometers north of Boulder, where the NBS Time and Frequency Division is located.

**RADIO TIME BROADCASTS**

1904 – UNITED STATES NAVAL OBSERVATORY TRANSMITS FROM BOSTON

1910 – TRANSMISSIONS FROM EIFFEL TOWER IN PARIS

1923 – NATIONAL BUREAU OF STANDARDS TRANSMITS FROM WASHINGTON D.C.

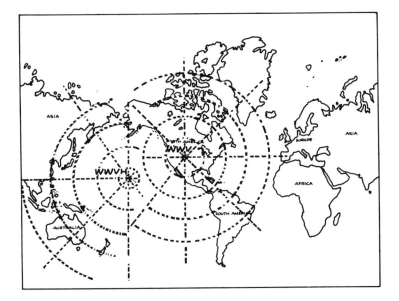

A sister station, WWVH, was installed in Maui, Hawaii, in 1948, to provide similar services in the Pacific area and western North America. In July 1971, WWVH was moved to a site near Kekaha on the Island of Kauai, in the western part of the Hawaiian Island chain. The 35 percent increase in area coverage achieved by installation of new and better equipment extended WWVH service to include Alaska, Australia, New Zealand, and Southeast Asia.

Throughout the years NBS has expanded and revised the services and format of its shortwave broadcasts to meet changing and more demanding needs. Today signals are broadcast at several different frequencies in the shortwave band 24 hours a day. The signal format provides a number of different kinds of information, such as standard musical pitch, standard time interval, a time signal both in the form of a voice announcement and a time code, information about radio broadcast conditions, and even weather information about major storm conditions in the Atlantic and Pacific areas from WWV and WWVH respectively.

The broadcasts of WWV may also be heard via telephone by dialing (303) 499-7111 (Boulder, Colorado). The telephone user will hear the live broadcasts as received by radio in Boulder. Considering the instabilities and variable delays of propagation by radio and telephone combined, the listener should not expect accuracy of the telephone time signals to better than $\frac{1}{30}$ of a second.

NBS also broadcasts a signal at 60 kHz from radio station WWVB, also located in Fort Collins, in the form of a time code, which is intended primarily for domestic uses. This station provides better quality frequency information because atmospheric propagation effects are relatively minor at 60 kHz (see Chapter 9). The time code is also better suited to applications where automatic equipment is utilized.

The USNO provides time and frequency information via a number of U. S. Navy communication stations, some of which operate in the very low frequency (VLF) range. The U. S. Navy has also been testing experimentally the possibility of disseminating time information from satellites, as has NBS (see page 141).

At present, there are over 30 different radio stations throughout the world that broadcast standard time and frequency signals. And as we shall see in the following discussions, other broadcasts —particularly those from radio navigation systems—are also sources of time and frequency information. But first let's discuss the basic characteristics of radio time signals.

### Accuracy

The basic limitation in shortwave radio transmission of time information is that the information received lacks the accuracy of the information broadcast. The signal broadcast takes a small but definite time to reach the listener; when the listener hears that the time is 9:00 A.M., it is really a very small fraction of a second after 9:00 A.M. If he knows how long it takes for the radio signal to reach him, he can allow for the delay and correct this reading accordingly. But where extremely accurate time information is needed, determining the delay precisely is a difficult problem because the signal does not normally travel in a direct line to the listener. Usually it comes to him by bouncing along a zig-zag path

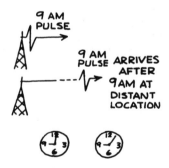

9 AM PULSE

9 AM PULSE ARRIVES AFTER 9 AM AT DISTANT LOCATION

between the surface of the earth and the ionosphere, which is a layer of the upper atmosphere that acts like a mirror for radio waves.

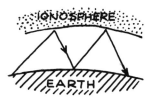

The height of this reflective layer depends in a complicated way upon the season of the year, the sunspot activity on the sun, the time of day, and many other subtle effects. So the height of the reflective layer changes constantly in a way that is not easy to predict, and thus the path delay in the signal is also difficult to predict or evaluate.

Because of these unpredictable effects, it is difficult to receive time by shortwave radio with an accuracy better than one one-thousandth of a second. For the everyday activities of about 98 percent of time information users, this degree of accuracy is more than adequate, of course. But there are many vitally important applications—such as the high-speed communications systems discussed in Chapter 11—where time must be known to one-millionth of a second, or even better.

## TIGER

The need for greater accuracy has led to several schemes for overcoming the problem of the unpredictable path delay. Instead of trying to predict or calculate the delay, for example, we can simply measure it. One of the most common ways to do this is to transmit a signal from the master clock, at a known instant of time, to the location we wish to synchronize. As soon as the signal is received at the remote location, it is transmitted back to the master clock. When the signal arrives back at its point of origin, we note its arrival time. Then by subtracting the time of transmission from the time of return, we can compute the round-trip time, and the one-way-trip time by dividing this figure by two.

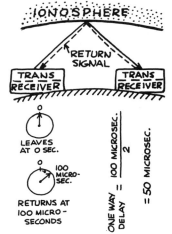

As is usually the case, however, we don't get something for nothing; we've had to install a transmitter at the receiver location to make the measurement. One of the most important applications of this particular approach has been to use a satellite to relay signals back and forth between the locations we wish to synchronize.

### Coverage

Accuracy is only one of the requirements for a usable time-information system. Obviously, an extremely accurate time source that could make its information known only a hundred miles away would be of limited use; sometimes it's important to make the same information simultaneously available almost throughout the world. During the International Geophysical Year, starting in July of 1957, for example, scientists wanted to know how certain geophysical events progressed as a function of time over the surface of the earth. They wanted to find out, among other things, how a large burst of energy from the sun affected radio communications on earth, as a function of time and location. Such information not only has practical importance for local and worldwide communications, but it also provides data to develop theories and to decide between proposed theories.

Unfortunately there was not—and still is not—any one system that provides adequate worldwide coverage. Different users have to use different systems with varying degrees of success. And there is always the important question of how accurately the various systems are tied to each other. Although it may be possible to recover the time from two different systems quite accurately, the end result is still no better than the degree to which stations are synchronized to each other.

### Reliability

Reliability is another important factor. Even if the transmitting station is almost never down because of technical difficulties, radio signals fade in and out for the receiver. Most of the well-known standard time and frequency broadcast services are in the shortwave band, where fading can be a severe problem. To return to our scientist of the Geophysical Year, he may want to make a crucial measurement during, say, an earthquake; and he discovers that there is no available radio time signal.

Of course, most users are aware of this difficulty; so they try to protect themselves against such loss of information by maintaining a clock at their own location, to interpolate between losses of radio signals. Or more usually they routinely calibrate their clocks once a day when reliable radio signals *are* available.

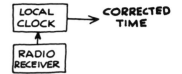

At the broadcast station an attempt is made to overcome loss of its signals by broadcasting the time on several different frequencies at once. The hope is that at least one signal will be available most of the time.

### Other Considerations

*Percent of time available* refers mainly to systems that are not particularly subject to signal fading but are off the air part of the time. For example, if we broadcast time over a commercial TV station, say, once every half hour, we can be pretty certain that the signal will be there when it is supposed to be. But since TV stations are off the air during late-night and early-morning hours, the information will not be available 100 percent of the time.

*Receiver cost* is another factor that the user must consider in his choice of system. No system is ideal for all users or in all circumstances. And as is generally the case in most choices, one has to accept some limitations in order to get the advantages most important to him.

*Ambiguity* refers to the degree to which the time signal is self-contained. For example, a time signal that consisted of ticks at one-minute intervals would allow the user to set the second hand of his watch to zero at the correct moment, but it wouldn't tell him at what minute to set the minute hand. On the other hand, if the minute ticks were preceded by a voice announcement that said, "At the tick the time is 12 minutes after the hour," then the listener could set both his second hand and his minute hand, but not his hour hand.

For the most part, shortwave radio broadcasts dedicated primarily to disseminating time are relatively unambiguous in the sense that they broadcast day, month, year, hour, minute, and second information. In some other services one assumes that the user knows what year, month, and day it is. Other systems, particularly navigation systems (see page 104), that are sources of time information, are usually more ambiguous because their signals are primarily ticks and tones; and the user must have access to some other time signal to remove the ambiguity.

## OTHER RADIO SCHEMES

In addition to the widely used shortwave broadcasts of time information, there exist other radio systems that can be used to retrieve such information. Low-frequency navigation systems, for instance, although they were built and are operated for another purpose, are good sources of time information because their signals are referenced to high-quality atomic frequency standards and to "official" time sources.

At the other end of the frequency spectrum, television broadcasts provide a source of extremely sharp, strong pulses that can readily be used to synchronize any number of clocks. Actually, *any* kind of radio signal with some identifiable feature that is "visible" at two or more places can be used to synchronize clocks, as we shall see in the next chapter. Of course, clocks can be synchronized without necessarily telling the "correct" or standard time. But if any one of the clocks being synchronized has access to standard time (date), all other clocks—given the necessary equipment—can be synchronized to it.

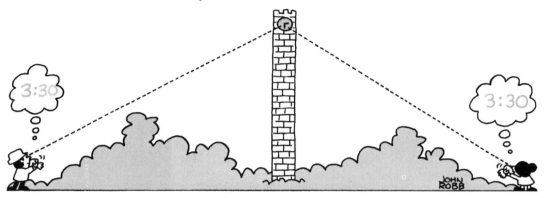

The broadcast *frequency* of a radio signal has a great deal to do with its usefulness. Systems other than the shortwave system have certain advantages—but also distinct disadvantages. We shall say more about this in the next chapter.

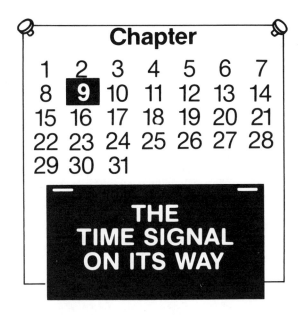

# Chapter

1 2 3 4 5 6 7
8 9 10 11 12 13 14
15 16 17 18 19 20 21
22 23 24 25 26 27 28
29 30 31

## THE TIME SIGNAL ON ITS WAY

We've already realized that information about the "correct time" is useless unless it's instantly available. But what, exactly, do we mean by "instantly," and how is it made available to everyone who wants to know what time it is? What can happen to this information on its way to the user, and what can be done to avoid some of the bad things that can and do happen?

## CHOOSING A RADIO FREQUENCY

The *frequency* of a radio signal primarily determines its *path*. The signal may bounce back and forth between the ionosphere and the surface of the earth, or creep along the curved surface of the earth, or travel in a straight line—depending on the frequency of the broadcast. We shall discuss the characteristics of different frequencies, beginning with the very low frequencies and working our way up to the higher frequencies.

### Very Low Frequencies (VLF)—3 - 30 kHz

The big advantage of VLF signals is that one relatively low-powered transmitter could provide worldwide coverage. A number of years ago, a VLF signal broadcast in the mountains near Boulder, Colorado, was detected in Australia, even though the broadcast signal strength was less than 100 watts. The VLF signal travels great distances because it bounces back and forth between the earth's surface and the lowest layer of the ionosphere, with very little of its energy being absorbed at each reflection.

Another good thing about VLF signals is that they are not strongly affected by irregularities in the ionosphere, which is not

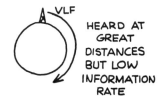

VLF HEARD AT GREAT DISTANCES BUT LOW INFORMATION RATE

true in the case of shortwave transmissions. This is so because the size of the irregularities in the ionosphere "mirror" is generally small compared to the length of the VLF radio waves. For example, at 20 kHz the radio frequency wavelength is 15 kilometers. The effect is somewhat the same as the unperturbed motion of a large ocean liner through slightly choppy seas.

But VLF also has serious limitations. One of the big problems is that VLF signals cannot carry very much information because the signal frequency is so low. We cannot, for example, broadcast a 100 kHz tone over a broadcast signal operating at 20 kHz. It would be like trying to get mail delivery ten times a day when the mailman comes only once a day. More practically, it means that time information must be broadcast at a very slow rate, and any schemes involving audio frequencies, such as voice announcement of the time, are not practical.

We mentioned that VLF signals are not particularly affected by irregularities in the ionosphere, so the path delay is relatively stable. Another important fact is that the ionospheric reflection height is about the same from one day to the next at the same time of day. Unfortunately, though, calculating the path delay at VLF is a complicated and tedious procedure.

One last curious thing about VLF is that the receiver is better off if he listens to signals from a distant station than from nearby stations. The reason is that near the station he gets two signals— one that is reflected from the ionosphere (sky wave) and another that is propagated along the ground. And what he receives is the sum of these two signals. This sum varies in a complicated way as a function of time and distance from the transmitter. So the listener wants to be so far from the transmitter that for all practical purposes the ground wave has died out, and he doesn't have to deal with this complicated interference pattern.

### Low Frequencies (LF)—30 - 300 kHz

In many respects LF signals have properties similar to VLF. Of course, the fact that the carrier frequencies are higher means that the information-carrying rate of the signal is potentially higher also.

These higher carrier frequencies have allowed the development of an interesting trick to improve the path stability of signals. The scheme was developed for the Loran-C navigation system (See pages 161-163) at 100 kHz, which is also used extensively to obtain time information. The trick is to send a pulsed signal instead of a continuous signal. A particular burst of signal will reach the observer by two different paths. He will first see the ground wave signal that travels along the surface of the earth. And a little later he will see the same burst of signal arriving via reflection from the ionosphere.

At 100 kHz, the ground wave arrives about 30 microseconds ahead of the ionospheric wave, and this is usually enough time to measure the ground wave, uncontaminated by the sky wave. The

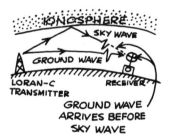

GROUND WAVE
ARRIVES BEFORE
SKY WAVE

ground wave is quite stable in path delay, and the path-delay prediction is considerably less complicated than when one is working with the sky wave.

Beyond about 1000 kilometers, however, the ground wave becomes so weak that the sky wave predominates; and at that point we are pretty much back to the kind of problems we had with VLF signals.

### Medium Frequencies (MF)—300 kHz - 3 MHz

We are most familiar with the medium-frequency band because it contains the AM broadcast stations of this country. During the day the ionospheric or sky wave is heavily absorbed, as it is not reflected back to earth; so for the most part, during the daytime we receive only the ground wave. At night, however, there is no appreciable absorption of the signal, and these signals can be heard at great distances.

One of the standard time and frequency signals is at 2.5 MHz. During the daytime, when the ground wave is available, the Japanese report obtaining 30-microsecond timing accuracy. At night, when one is receiving the sky wave, a few milliseconds is about the limit. It is fair to say, however, that this band has not received a great deal of attention for time dissemination, and there may be future promise here.

### High Frequencies (HF)—3.0 - 30 MHz

The HF band is the one we usually think of when we speak of the shortwave band. Signals in this region are generally not heavily absorbed in reflection from the ionosphere. Absorption becomes even less severe as we move toward the upper end of this band. Thus the signal may be heard at great distances from a transmitter, but it may arrive after many reflections, so accurate delay prediction is difficult.

Another difficulty is that in contrast to VLF waves, HF wave lengths can be of the same order, or smaller, than irregularities in the ionosphere. And since these irregularities are constantly changing their shape and moving around, the signal strength at a particular point will fade in and out in amplitude. Because of the fading and the continuous change in path delay, accuracy in timing is again restricted to about 1 millisecond, unless one is near enough to the transmitter to receive the ground wave.

Most of the world's well-known standard time and frequency broadcasts are in this band.

### Very High Frequencies (VHF)—30 - 300 MHz

From a propagation point of view, one of the most important things that happen in the VHF band is that the signals are often not reflected back to the surface of the earth, but penetrate through the ionosphere and propagate to outer space. This means that we can receive only those stations that are in line of sight, and explains why we do not normally receive distant TV stations, TV signals being in this band. It also means that many different

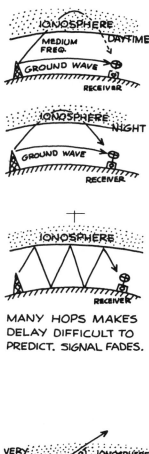

MANY HOPS MAKES DELAY DIFFICULT TO PREDICT. SIGNAL FADES.

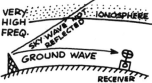

LIMITED TO LINE OF SIGHT. HAS HIGH ACCURACY. CAN SEND SHARP PULSES.

signals can be put on the same channel, and there is little chance of interference as long as the stations are separated by 300 kilometers or more.

From a timing point of view, however, this is bad; for if we want to provide worldwide—or even fairly broad—coverage, many stations are required, and they must all be synchronized. On the other hand, there are advantages to having no ionospheric signals because this means that we can receive a signal uncontaminated by sky wave. We can also expect that once we know the delay for a particular path, it will remain relatively stable from day to day.

A third advantage is that because carrier frequencies are so high, we can send very sharp rise-time pulses, and thus can measure the arrival time of the signals very precisely. Because of the sharp rise time of signals and the path stability, timing accuracies in this region are very good. Microsecond timing is relatively easy; and if some care is taken, even 0.1 microsecond can be achieved.

At the time of this writing the master clock at NBS in Boulder, Colorado, is used as a reference clock for the standard time and frequency broadcasts from WWV near Fort Collins, some 80 kilometers away. A TV signal is used to maintain the link between Boulder and Fort Collins with an accuracy of a small fraction of a microsecond. The system is explained more fully at the end of the next chapter.

### Frequencies Above 300 MHz

The main characteristic of these frequencies is that like VHF frequencies the signals penetrate into outer space, so systems are limited to line-of-sight. There may be problems caused by small irregularities in the path—or "diffraction effects," as they are commonly called—similar to such effects at optical wavelengths. Nevertheless, if a straight shot to the transmitter is available, we can expect good results.

Above 1000 MHz weather may produce problems; this is especially significant in broadcasting time from a satellite, where we wish to minimize both ionospheric and lower atmosphere effects.

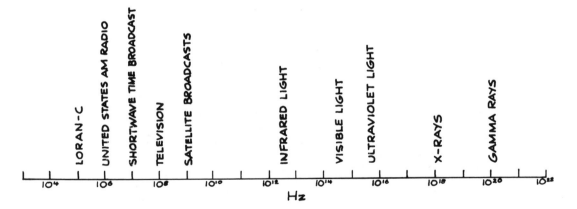

## NOISE—ADDITIVE AND MULTIPLICATIVE

We have been discussing different effects that we should expect in the various frequency bands. We should differentiate explicitly here between two types of effects on the signals. They are called "additive noise" and "multiplicative noise." "Noise" is the general term used to describe any kind of interference that mingles with or distorts the signal transmitted and so contaminates it.

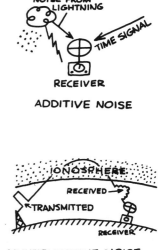

ADDITIVE NOISE

Additive noise is practically self-explanatory. It refers to noise added to the signal that reduces its usefulness. For example, if we were listening to a time signal that was perturbed by radio noise caused by lightning or automobile ignition noise, we would have an additive noise problem.

Multiplicative noise is noise in the sense that something happens to the signal to distort it. A simple illustration is the distortion of one's image by a mirror in a fun house. None of the light or signal is *lost*—it is just rearranged so that the original image is *distorted*. The same thing can happen to a signal as it is reflected by the ionosphere. What was transmitted as a nice, clean pulse may, by the time it reaches the user, be smeared out or distorted in some way. The amount of energy in the pulse is the same as if it had arrived cleanly, but it has been rearranged.

MULTIPLICATIVE NOISE

How can we overcome noise? With additive noise, the most obvious thing to do is to increase the transmitter power so that the received signal-to-noise ratio is improved. Another way is to divide the available energy and transmit it on several different frequencies at once. It may be that one of these frequencies is extremely free from additive noise. Another possibility—quite often used—is to "average" the signal. We can take a number of observations, average them, and improve our result. This works because the information on the signals is nearly the same all the time, so the signal keeps building up; but the noise is, in general, different from one instant to the next; therefore it tends to cancel itself out.

With multiplicative noise it doesn't help to increase the transmitter power. To return to our previous illustration, the image from a fun-house mirror will be just as *distorted* whether the viewer is standing in bright light or in dim light. Most of the strategies for overcoming multiplicative noise come under the general heading of *diversity*—specifically space diversity, frequency diversity, and time diversity.

- Space diversity means that we measure the incoming signal at several different *locations,* but not just any location. The locations must be far enough apart so that we are not seeing the same distortion; what we are attempting to do is to look for elements in the signal that are common to all signals. In other words, if we look at the fun-house mirror from several different locations, it is the distortion that changes, not our body. And maybe by looking at the mirror from several different locations we can get the distortions to cancel out, leaving the true image.
- Frequency diversity means simply that we send the same information on several different *frequencies,* again hoping that the signal distortion on the different frequencies will be sufficiently different so that we can cancel them out and obtain the true signal image.
- Time diversiity means that we send the same message at different *times,* hoping that the distortion mechanism will have changed sufficiently between transmissions for us to reconstruct the original signal.

## THREE KINDS OF TIME SIGNALS

There are basically three different kinds of signals that we can use to get time information. The most obvious is, of course, a signal that was constructed for this very purpose, such as a broadcast from the National Bureau of Standards (NBS) shortwave station WWV, or perhaps the time announcement on the telephone. The obvious utility of this method is that the information comes to us in a relatively straightforward way, and we have to do very little processing.

A second way is to listen to some signal that has time information buried in it. A good example is the Loran-C navigation system. In this system the pulses emitted for navigation are related in a very precise way to atomic clocks that are all coordinated throughout the system. Although we may not get a pulse exactly on the second, minute, or hour, the emission times of these signals are related precisely to the second, minute, and hour. So to use Loran-C for time information we must, of course, make a measurement of the arrival time of a pulse or pulses with respect to our own clock. And we must also have information that tells us

STANDARD TIME
BROADCASTS

RADIO NAVIGATION
SIGNALS

TV SIGNALS

how the pulses are related to the controlling clock. This information is, in fact, available about Loran-C from the United States Naval Observatory (USNO).

Finally, we can use a radio signal for synchronization without any specific effort by those operating the transmitter to provide such information—or even knowing that it is being so used. The process is called the "transfer standard" technique; and it is used, for example, to keep the radio emissions from the standard time-and-frequency shortwave radio stations at Fort Collins, Colorado, referenced back to the atomic clock system and National Frequency Standard at the National Bureau of Standards Laboratories 80 kilometers away, in Boulder.

Let's see how the method works. A TV signal consists of a number of short signals in quick succession, with each short signal responsible for one line of the picture on the TV screen. Each such signal is about 63 microseconds long, and each is preceded by a pulse that, in effect, tells the TV set to get ready for the next line of information. Let's suppose now that we recorded the arrival time of one of these get-ready pulses—or "synchronization pulses," as they are called—with respect to our clock in Boulder. Let's suppose that someone at the station in Fort Collins also monitors the arrival time of this same pulse with respect to the clock at Fort Collins.

The TV stations in this area are near Denver, which is closer to Boulder than to Fort Collins. So we will see a particular "sync pulse" in Boulder before it is seen in Fort Collins, because of the extra distance it must travel to reach Fort Collins. If we assume that we have measured or calculated this extra path delay, we see that we are in a position now to check the clocks in Boulder and Fort Collins against each other.

It could work like this: The man who made the measurement in Boulder could call his fellow measurer in Fort Collins, and they could compare readings. If the two clocks are synchronized, then the Boulder time-of-arrival measurement subtracted from the Fort Collins time-of-arrival measurement should equal the extra path delay between Boulder and Fort Collins. If the measurement is either greater or smaller, then the two clocks are out of synchronization, and by a known amount that can be found simply by subtracting the known path delay from the measured difference.

We have seen that any kind of radio signal with some identifiable feature that is visible at two or more places can be used in this way to synchronize clocks. As we've mentioned, though, a TV signal provides a remarkably sharp, clear signal, free from problems inherent in any system that bounces signals off the unpredictable ionosphere. But of course its coverage is limited to a radius of about 200 miles from the TV station.

Finding out what time it is can be very simple, or very complicated, depending on *where* one is when he needs to know the time, and how accurate a time he needs to know.

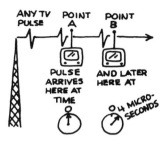

FROM PREVIOUS MEASUREMENTS OPERATORS AT A AND B KNOW THAT IT TAKES A SIGNAL FOUR MICROSECONDS TO TRAVEL FROM POINT A TO POINT B. TO CHECK THE SYNCHRONIZATION OF THEIR CLOCKS, EACH RECORDS THE ARRIVAL TIME OF THE SAME PULSE AS IT ARRIVES FIRST AT CLOCK A AND THEN AT CLOCK B. IF THE CLOCKS ARE SYNCHRONIZED THEIR TWO ARRIVAL TIME READINGS WILL DIFFER BY 4 MICROSECONDS. IF NOT, THEY CAN USE THE DISCREPANCY TO DETERMINE THE AMOUNT OF SYNCHRONIZATION ERROR BETWEEN THE TWO CLOCKS.

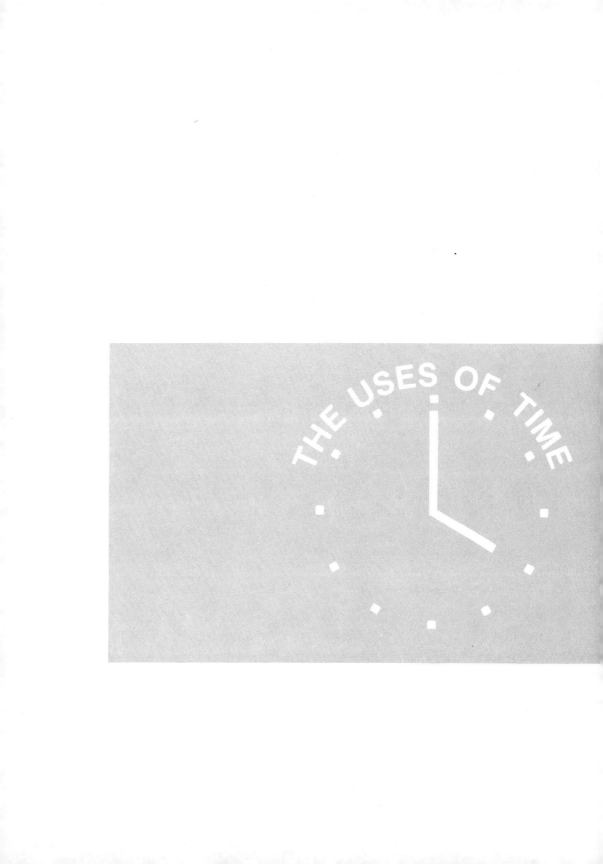

THE USES OF TIME

# IV
# THE USES OF TIME

# Chapter

| 1 | 2 | 3 | 4 | 5 | 6 | 7 |
| 8 | 9 | **10** | 11 | 12 | 13 | 14 |
| 15 | 16 | 17 | 18 | 19 | 20 | 21 |
| 22 | 23 | 24 | 25 | 26 | 27 | 28 |
| 29 | 30 | 31 | | | | |

## STANDARD TIME

Our discussions of time have shown that "the time" when something happened or the "length" of time it lasted depends on the scale used for measurement. "Sun time" is different from "star time," and both differ from "atomic time." Sun time at one location is inevitably different from sun time only a few kilometers—or even a few *meters*—east or west.

A factory whistle blown at 7:00 in the morning, noon, 1:00 and 4:00 P.M. to tell workers when to start and stop their day's activities served many a community as its time standard for many years. It didn't matter that "the time" was different in each community. But in today's complex society, with its national and international networks of travel and communications systems, it's obvious that some sort of universal *standard* is essential. The establishment of *Standard Time* is much more recent than many persons realize.

### STANDARD TIME ZONES AND DAYLIGHT-SAVING TIME

In the latter part of the 19th century a traveler standing in a busy railroad station could set his pocket watch to any one of a number of clocks on the station wall; each clock indicated the "railroad time" for its own particular line. In some states there were literally dozens of different "official" times—usually one for each major city—and on a cross-country railroad trip the traveler would have to change his watch 20 times or so to stay in step with the "railroad time." It was the railroads and their pressing need for accurate, uniform time, more than anything else, that led to the establishment of *time zones* and standard time.

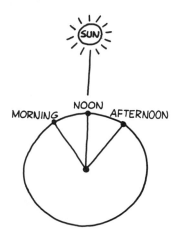

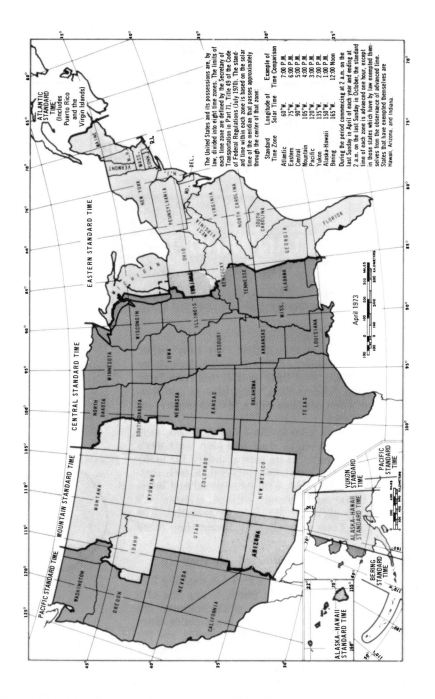

The United States and its possessions are, by law, divided into eight time zones. The limits of each time zone are defined by the Secretary of Transportation in Part 71, Title 49 of the Code of Federal Regulations (July 1970). The standard time within each zone is based on the solar time of the meridian that passes approximately through the center of that zone:

| Standard Time Zone | Longitude of Solar Time | Example of Time Comparison |
|---|---|---|
| Atlantic | 60°W. | 7:00 P.M. |
| Eastern | 75°W. | 6:00 P.M. |
| Central | 90°W. | 5:00 P.M. |
| Mountain | 105°W. | 4:00 P.M. |
| Pacific | 120°W. | 3:00 P.M. |
| Yukon | 135°W. | 2:00 P.M. |
| Alaska-Hawaii | 150°W. | 1:00 P.M. |
| Bering | 165°W. | 12:00 Noon |

During the period commencing at 2 a.m. on the last Sunday in April of each year and ending at 2 a.m. on the last Sunday in October, the standard time of each zone is advanced one hour, except in those states which have by law exempted themselves from the observance of advanced time. States that have exempted themselves are Hawaii, Arizona, and Indiana.

April 1973

STANDARD TIME ZONES OF THE UNITED STATES OF AMERICA

One of the early advocates of uniform time was a Connecticut school teacher, Charles Ferdinand Dowd. Dowd lectured railroad officials—and anyone else who would listen—on the need for a standardized time system. Since the continental United States covers approximately 60 degrees of longitude, Dowd proposed that the nation be divided into four zones, each 15 degrees wide—which is the distance the sun travels in one hour. With the prodding of Dowd and others, the railroads adopted in 1883 a plan that provided for five time zones—four in the United States and a fifth covering the easternmost provinces of Canada.

The plan was placed in operation on November 18, 1883. There was a great deal of criticism. Some newspapers attacked the plan on the grounds that the railroads were "taking over the job of the sun," and said that, in fact, the whole world would be "at the mercy of railroad time." Farmers and others predicted all sorts of dire results—from the production of less milk and fewer eggs to drastic changes in the climate and weather—if "natural" time was interfered with. Local governments resented having their own time taken over by some outside authority. And so the idea of a standard time and time zones did not gain popularity rapidly.

But toward the end of the second decade of the 20th century the United States was deeply involved in a World War. On March 19, 1918, the United States Congress passed the Standard Time Act, which authorized the Interstate Commerce Commission to establish standard time zones within the United States; and at the same time the Act established "daylight-saving time," to save fuel and to promote other economies in a country at war.

The United States, excluding Alaska and Hawaii, is divided into four times zones; the boundary between zones zigzags back and forth in a generally north-south direction. Today, for the most part, the time-zone system is accepted with little thought, although some people near the boundaries still complain and even gain boundary changes so that their cities and towns are not "unnaturally" separated from neighboring geographical regions where they trade or do business.

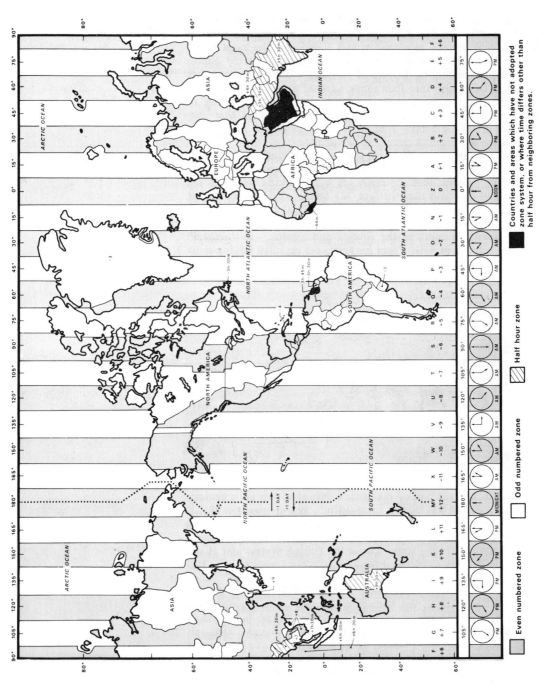

**STANDARD TIME ZONES OF THE WORLD**

The idea of "daylight-saving time" has roused the emotions of both supporters and critics—notably farmers, persons responsible for transportation and radio and television schedules, and persons in the evening entertainment business—and continues to do so. Rules governing daylight-saving time have undergone considerable modification in recent years. Because of confusion caused by the fact that some cities or states chose to shift to daylight-saving time in summer and others did not—with even the dates for the shifts varying from one place to another—Congress ruled in the Uniform Time Act of 1966 that the entire nation should use daylight-saving time from 2:00 A.M. on the last Sunday in April until 2:00 A.M. on the last Sunday in October. (Actually "daylight-saving time" does not exist: there is only "standard time" which is advanced one hour in the summer months. "Daylight-saving time" has no legal definition, only a popular understanding.) Any *state* that did not want to conform could, by legislative action, stay on standard time. Hawaii did so in 1967, Arizona in 1968—(though Indian reservations in Arizona—which are under Federal jurisdiction—use daylight-saving time) and Indiana in 1971. In a 1972 amendment to the Uniform Time Act those states split by time zones may choose to keep standard time in one part of the state and daylight-saving time in the other. Indiana has taken advantage of this amendment so that only the western part of the state observes daylight-saving time.

When fuel and energy shortages became acute in 1974, it was suggested that a shift to daylight-saving time throughout the nation the year around would help to conserve these resources. But when children in some northern areas had to start to school in the dark in winter months, and the energy savings during these months proved to be insignificant, year-around daylight-saving time was abandoned, and the shifts were returned to the dates originally stated by the 1966 Uniform Time Act. In the long run the important thing is that the changes be uniform and that they apply throughout the nation, as nearly as possible.

The whole world is divided into 24 standard time zones, each approximately 15 degrees wide in longitude. The zero zone is centered on a line running north and south through Greenwich, England. The zones to the east of Greenwich have time later than Greenwich time, and the zones to the west have earlier times—one hour difference for each zone.

With this sytem it is possible for a traveler to gain or lose a day when he crosses the International Date Line, which runs north and south through the middle of the Pacific Ocean, 180° around the world from Greenwich. A traveler crossing the line from east to west automatically advances a day, whereas one traveling in the opposite direction "loses" a day.

Both daylight-saving time and the date line have caused a great deal of consternation. Bankers worry about lost interest, and law suits have been argued and settled—often to no one's satisfaction—on the basis of whether a lapsed insurance policy covered substantial loss by fire, since the policy was issued on standard time and the fire in question, had it occurred during a period of standard instead of daylight-saving time, would have been within the time still covered by the policy. The birth or death date affected by an individual's crossing the date line can have important bearing on anything from the child's qualifying for age requirements to enter kindergarten to the death benefits to which the family of the deceased are entitled. The subject continues to be a lively issue, and probably will remain so.

## TIME AS A STANDARD

TIME INTERVAL—LOCAL

SYNCHRONIZATION—
     REGIONAL

DATE — UNIVERSAL

The disarray in railroad travel caused by the lack of a standard time system in the late 19th century illustrates one of the primary benefits of standardization—standards promote better understanding and communication. If we agree on a particular standard of time or mass, then we all know what a "minute" or a "kilogram" means.

In working with time and frequency, we have standardization at various levels. With the development of better clocks, people began to see the need for defining more carefully the basic units of time—since the minutes or seconds yielded by one clock were measurably different from those yielded by another. As early as 1820, the French defined the second as "1/86,400 of the mean solar day," establishing a standard time *interval* even though town clocks ticking at the same rate would show different local time—that is, a different *date* for each town.

In our first chapter we discussed briefly the concepts of *time interval, synchronization,* and *date.* In a sense these three concepts represent different levels of standardization. Time interval has a kind of "local" flavor. When one is boiling a three-minute egg, the time in Tokyo is of little concern to him. What he needs to know is how long three minutes is at his location.

## TIGER

©KING FEATURES SYNDICATE,INC.1977—REPRINTED BY PERMISSION OF KING FEATURES SYNDICATE

Synchronization has a somewhat more cosmopolitan flavor. Typically, if we are interested in synchronization we care only that particular events start or stop at the same time, or that they stay in step. For example, if people on a bus tour are told to meet at the bus at 6:00 P.M., they need only synchronize their watches with the bus driver's watch, to avoid missing the bus. It is of little consequence whether the bus driver's watch is "correct" or not.

The concept of date has the most nearly universal flavor. It is determined according to well-defined rules discussed on page 67, and it cannot be arbitrarily altered by people on bus tours; they do it only at their own peril, for they may well be late for dinner.

There has been a trend in recent years to develop standards in such a way that, if certain procedures are followed, the basic units can be determined. For example, the definition of the second, today, is based upon counting a precise number of oscillations of the cesium atom, as we discussed on page 66. This means that anybody who has the means and materials necessary, and who is clever enough to build a device to count vibrations of the cesium atoms, can determine the second. He doesn't have to travel to Paris. Similarly, the unit of length is defined by a certain wave length of light emitted by the krypton atom.

Concepts such as date, on the other hand—built from the basic units—have an arbitrary starting point—such as the birth of Christ—which cannot be determined by any physical device.

## IS A SECOND REALLY A SECOND?

In our development of the history of timekeeping, we saw that the spinning earth makes a very good timepiece; even today, except for the most precise needs, it is more than adequate. Nevertheless, with the development of atomic clocks we have turned away from the earth definition of the second to the atomic definition. But how do we know that the atomic second is uniform?

One thing we might do to find out is to build several atomic clocks, and check to see if the seconds they generate "side by side" are of equal length. If they are, then we will be pretty certain that we can build clocks that produce uniform intervals of time at the "same" time.

But then how can we be certain that the atomic second itself isn't getting longer or shorter with time? Actually there is no way to tell, if we are simply comparing one atomic clock with another. We must compare the atomic second with some other *kind* of second. But then if we measure a difference, which second is changing length and which one is not? There would seem to be no way out of this maze. We must take another approach.

Instead of trying to prove that a particular kind of clock produces uniform time, the best we can do is agree to take some device—be it the spinning earth, a pendulum, or an atomic clock —and simply say that the output of that device helps us define time. In this sense we see that time is really the result of some set of operations that we agree to perform in the same way. This set of operations produces the standard of time; other sets of operations will produce different time scales.

But, one may ask, what if our time standard really does speed up at certain times and slow down at others? The answer is that it really doesn't make any difference, because all clocks built on the same set of operations will speed up and slow down together, so "we will all meet for lunch at the same time"—it's a matter of definition.

## WHO CARES ABOUT THE TIME?

Every day hundreds of thousands of people drop nickels, dimes, and quarters into parking meters, coin-operated washers, dryers and dry-cleaning machines, and "fun" machines that give their children a ride in a miniature airplane or on a mechanical horse. Housewives trust their cakes and roasts, their clothing and their fine china to timers on ovens, laundry equipment, and dishwashers. Businesses pay thousands of dollars for the use of a computer's time or for minutes—sometimes fractions of minutes—of a communication system's time. We all pay telephone bills based on the number of minutes and parts of minutes we spend talking to Aunt Martha halfway across the nation.

The pumps at the gas station and the scales at the supermarket bear a seal that certifies recent inspection by a standards authority, and assurance that the device is within the accuracy requirements set by law. But who cares about the devices that measure time? What's to prevent a company from manufacturing equipment that runs for 9 minutes and 10 seconds, for instance, instead of the 10 minutes stated on a label? Are there any regulations at all for such things?

Yes indeed. In the United States, the National Bureau of Standards (NBS) has the responsibility for developing and operating standards of time interval (frequency). It is also given the responsibility of providing the "means and methods for making measurements consistent with those standards." As a consequence of these directives, the NBS maintains, develops, and operates a primary frequency standard based on the cesium atom. It also broadcasts standard frequencies based on this primary standard. (See page 73.)

The state and local offices of weights and measures deal with matters of time interval and date, generally by reference to an NBS handbook that deals with such devices as parking meters, parking garage clocks, "time in-time out" clocks, and similar timing devices. The greatest accuracies involved in these devices are about ± 2 minutes on the date, and about 0.1 percent on time interval. Typically the penalty for violating this code is a fine, a jail sentence, or both, for the first offense.

State standards laboratories seek help from NBS for such duties as calibrating radar "speed guns" used by traffic officers and other devices requiring precise timing. In addition to NBS, there are more than 250 commercial, governmental, and educational institutions in the United States that maintain standards laboratories; some 65 percent of these do frequency and/or time calibrations. So the facilities for monitoring the timing devices that affect the lives of all of us are readily available throughout the land.

In the United States, the United States Naval Observatory (USNO) collects astronomical data essential for safe navigation at sea, in the air, and in space. The USNO maintains an atomic time scale based on a large number of commercial cesium-beam frequency standards. And like NBS, it disseminates its standard, or time scale, by providing time information to several U.S. Navy broadcast stations. The Department of Defense (DOD) has given the USNO the responsibility of tending to the time and frequency needs of the DOD. As a practical matter, however, both the USNO and NBS have a long history of working cooperatively together to meet the needs of a myriad of users.

The responsibility for enforcing the daylight-saving time changes and keeping track of the standard time zones in this country is held by the U.S. Department of Transportation (DOT). And yet another organization—the Federal Communications Commission (FCC)—is involved in time and frequency control through its regulation of radio and television broadcasts. Its *Code of Federal Regulations—Radio Broadcast Services* describes the frequency allocations and the frequency tolerances to which various broadcasters must conform. These include AM stations, commercial and non-commercial FM stations, TV stations, and international broadcasts. The NBS broadcast stations are references which the broadcaster may use to maintain assigned frequency, but the FCC is the enforcing agency.

The development, establishment, maintenance, and dissemination of information generated by time and frequency standards are vitally important services that most of us take for granted and rarely question or think about at all; and they require constant monitoring, testing, comparisons, and adjustment. Those responsible for maintaining these delicate and sensitive standards are constantly seeking better ways to make them more widely available, at less cost to more users. Each year, the demand for better, more reliable, and easier-to-use standards grows; and each year the scientists come up with at least some new concepts and answers to their problems.

**Chapter**

1 2 3 4 5 6 7
8 9 10 **11** 12 13 14
15 16 17 18 19 20 21
22 23 24 25 26 27 28
29 30 31

# TIME, THE GREAT ORGANIZER

Time is so basic a part of our daily lives that we tend to take it for granted and overlook the vital part it plays in industry, scientific research, and many other activities of our present-day world. Almost any activity today that requires precision control and organization rests on time and frequency technology. Its role in these activities is essentially the same as in our own mundane affairs—providing a convenient way to bring order and organization into what would otherwise be a chaotic world.

The difference is mainly one of degree; in our everyday lives we rarely need time information finer or more accurate than a minute or two, but modern electronics systems and machines often require accuracies of one microsecond and better. In this chapter we shall see how the application of precise time and frequency technology helps to solve problems of control and distribution in three key areas of modern industrial society—*energy, communication,* and *transportation*—plus a few other usual and unusual uses of time and frequency information.

ENERGY

COMMUNICATION

TRANSPORTATION

## ELECTRIC POWER

Whether it is generated by nuclear reactors, fossil fuel-burning plants, or a hydroelectric system, electric power is delivered in the United States and Canada at 60 Hz, and at 50 Hz in a good part of the rest of the world. For most of us it is in this aspect of electric power that time and frequency plays its most familiar role. The kitchen wall clock is not only powered by electricity, but its "ticking" rate is tied to the "line" frequency maintained by the power company.

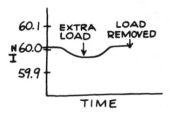

The power companies carefully regulate line frequency, so electric clocks keep very good time. The motors that drive tape and record players operate at rates controlled by the line frequency, so that listeners hear the true sound; and electric toothbrushes and shavers, vacuum cleaners, refrigerators, washing and drying machines operate efficiently.

Nevertheless, there are slight variations in frequency that cannot be overcome. If the line load unexpectedly increases in a particular location—such as when many people turn on their television sets at the same time to see a local news flash—power generators in the area will slow down until input energy to the system is increased or until the load is removed. For example, the line frequency may drop to 59.9 Hz for a time and then return to 60.0 Hz when the extra load is removed.

During the period when the line frequency is low, electric clocks will accumulate a time error that remains even though 60 Hz is restored at some later point. To remove this time error, it is the practice of the power companies to increase line frequency above 60 Hz until the time error is removed—at which time they drop back to 60 Hz. Generally the time error never exceeds two seconds; and in the United States this error is determined with respect to special time and frequency broadcasts of the National Bureau of Standards. (See page 74.)

But frequency plays a greater role in electric power systems than merely providing a convenient time base for electric clocks. Frequency is a basic quantity that can be measured easily at every point of the system, and thus provides a way to "take the temperature" of the system.

We have seen that frequency excursions are indicative of load variations in power consumption. These variations are used to generate signals that control the supply of energy to the generators, usually in the form of steam—or water at hydroelectric plants. To provide more reliable service, many power companies have formed regional "pools," so that if power demands in a particular region exceed local capability, neighboring companies can fill in with their excess capacity.

Frequency plays an important part in these interconnected systems from several standpoints. First, all electric power in a connected region must be at the same frequency. If an "idle" generator is started up to provide additional power, it must be running in synchronism with the rest of the system before being connected into it. If it is running too slowly, current will flow into its windings from the rest of the system in an attempt to bring it up to speed; and if it is running too fast, excessive current will flow out of its windings in an attempt to slow it down. In either case, these currents may damage the machine.

Besides running at the same electrical frequency as the rest of the system the new generator must also be in step, or in *phase* with the rest of the system. Otherwise, damaging currents may again flow to try to bring the machine into phase.

We can understand the distinction between phase and frequency by considering a column of soldiers marching to a drummer's beat. If the soldiers step in time to the beat, they will all be walking at the same rate, or *frequency;* but they will not be in *phase* unless all left feet move forward together. Power companies have developed devices that allow them to make certain that a new generator is connected to a system only after it is running with the correct frequency and phase.

In power pools, frequency helps to monitor and control the distribution and generation of power. Based on customer demand and the efficiency of various generating components in the system, members of power pools have developed complex formulas for delivering and receiving electric power from one another. But there are unexpected demands and disruptions in the system—a fallen line, for example—that produce alterations of these schedules. To meet scheduled as well as unscheduled demands, electric power operators use a control system that is responsive both to electric energy flow between neighboring members of a pool and to variations in system frequency. The net result of this approach is that variations in both frequency and scheduled deliveries of power are minimized.

Time and frequency technology is also a helpful tool for fault location—such as a power pole toppled in a high wind. The system works somewhat like the radio navigation systems described later in this chapter. (See page 104.) At the point in the distribution system where the fault occurs, a surge of electric current will flow through the intact lines and be recorded at several monitoring stations. By comparing the relative arrival time of a particular surge as recorded by the monitors, operators can determine the location of the fault.

The East Coast blackout a few years ago brought to attention the vital role that coordination and control—or the lack of them—play in the delivery of reliable electric power. Today many power companies are developing better and more reliable control systems. One of the requirements of such better systems will be to gather more detailed information about the system—information such as power flow, voltage, frequency, phase, and so forth—which will be fed into a computer for analysis. Much of this information will have to be carefully gathered as a function of *time*, so that the evolution of the power distribution system can be carefully monitored. Some members of the industry suggest that time to 50 microseconds and even better will be required in future control systems.

## MODERN COMMUNICATION SYSTEMS

Time and frequency technology is, if anything, even more basic and vital to the operation of communication systems than it is to the control of electric power systems. Time and frequency information is used to help keep track of messages and to make certain that they reach their intended destination.

One of the most familiar applications of frequency for message identification is "tuning" our radio or TV set to a desired station. What we are really doing is telling the set to select the correct frequency for the station we wish to tune in. When we turn the dial to channel 5, for example, the TV set internally "tunes" to the frequency that is the same as channel 5's broadcast frequency; the set thus selects just one specific frequency from all of those coming in, displaying it and screening out all others.

Of course, channel 5 will have a number of different programs throughout the day, and we may be interested in watching only the program scheduled for 8:00 P.M. We select this program by consulting our watch or clock. Thus we use frequency information to help us select the correct channel, and time information to help us select the desired program on the particular channel.

This use of time and frequency information is routine, but there are other, newer kinds of communication systems that make heavier demands on time and frequency technology. Let's consider a communication system which will provide eight distinct message channels. We might use these channels to connect eight pairs of people—each pair consisting of a person at the "send" end of the communication link and another at the "receive" end. At the send

end we have a device which scans each of the eight channels in a round-robin fashion. At any particular instant, only the message from one channel is leaving the scanning device, but during the time that it takes the pointer of the scanner to make one complete revolution, the output signal will be made of the parts of eight different messages interleaved together. The interleaved messages travel down some communication link, such as a telephone line, and are then fed into another scanning device which sorts the interleaved messages into their original forms. As the sketch shows, the scanning devices at the two ends of the communication link must be synchronized. If they are not, the messages leaving the scanning device on the receive end will be garbled. In some very high-speed communication systems the scanning devices must be synchronized to a few microseconds. This kind of communication system where the signals are divided into time slots is called "time division multiplexing."

As another possibility, we could send the eight messages simultaneously, but at eight different frequencies. Now we must know which *frequency* to tune to. This kind of scheme is called "frequency division multiplexing." Many systems combine time and frequency division multiplexing so that the senders and users must have clocks that synchronize in both time and frequency.

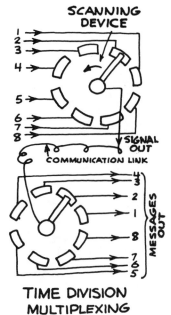

**TIME DIVISION MULTIPLEXING**

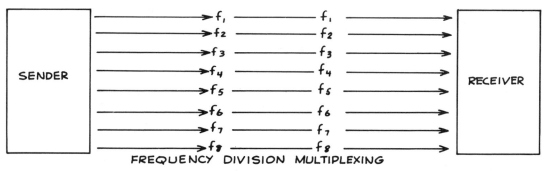

**FREQUENCY DIVISION MULTIPLEXING**

Since no clocks are perfect, they will gradually drift away from each other. So it will occasionally be necessary to use the communication system itself to make certain that all of the clocks involved show the same time.

One of the ways this can be done is for one of the users in the system to send a pulse that leaves his location at some particular time—say 4:00 P.M. Another user at a different location notes the arrival time of the 4:00 P.M. signal with respect to his clock. The signal should arrive at a later time, which is exactly equal to the delay time of the signal. If the listener at the receiving end records a signal that is either in advance of or after the correct delay time, he will know that the sender's clock has drifted ahead of or behind his own clock, depending upon the arrival time of the signal.

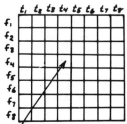

EXAMPLE: THIS MESSAGE IS TRANSMITTED AT TIME $t_4$ AT FREQUENCY $f_4$

**TIME AND FREQUENCY DIVISION MULTIPLEXING**

By expanding on this scheme all of the clocks in a system can be readjusted to the same time, or synchronized, simply by use of a synchronization pulse every so often. How often the readjustment must be made depends upon the quality of the clocks in the system, and also the rate at which the information is delivered. In a very familiar example—the television broadcast that we receive in our homes—there are about 15,000 synchronization pulses every second, a few percent of the communication capacity of the system. We shall discuss this example more fully in Chapter 16.

If we want to optimize the amount of time that the system is used to deliver messages, and minimize the amount of time devoted simply to the bookkeeping of synchronizing the clocks, then we must use the very best clocks available. This is one big reason for the continuing effort to produce better clocks and better ways of disseminating their information.

## TRANSPORTATION

In the second chapter of this book we discussed the important part time plays in navigation by the stars. But time is also an important ingredient in modern electronic navigation systems, in which the stars have been replaced by radio beacons.

Just as a road map is a practical necessity to a cross-country automobile trip, so airplanes and ships need their "road maps" too. But in the skies and on the oceans there are very few recognizable "sign-posts" to which the traveler can refer. So some artificial sign-post system has to be provided. In the early days of sailing vessels, fog horns, buoys, and other mechanical guides were used. Rotating beacon lights have long been used to guide both air and sea travel at night. But their ranges are comparatively short, particularly in cloudy or foggy weather.

Radio beacons seem to be the answer. Radio waves can be detected almost at once at long distance, and they are little affected by inclement weather. The need for long-range, high-accuracy radio navigation systems became critical during World War II. Celestial navigation and light beacons were virtually useless for aircraft and ships, especially in the North Atlantic during wintertime fog and foul weather. But time and frequency technology, along with radio signals, helped to provide some answers by constructing reliable artificial sign posts for air, sea, and even land travelers.

### Navigation by Radio Beacons

To understand the operation of modern radio navigation systems, let's begin by considering a somewhat artificial situation. Let's suppose we are on a ship located at exactly the same distance from three different radio stations, all of which are at this moment broadcasting a noontime signal. Because radio waves do not travel at infinite speed, the captain of our ship will receive the three noon signals a little past noon, but all at the same time. This simultaneous arrival of the time signals tells him that his ship is the same distance from each of the three radio stations.

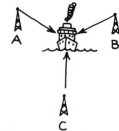

SIGNALS LEAVE ALL 3 TRANSMITTERS AT 12 NOON AND ARRIVE SIMULTANEOUSLY SHORTLY PAST NOON AT SHIP

If the locations of the radio stations are indicated on the captain's nautical map, he can quickly determine his position. If he were a little closer to one of the stations than he is to the other two, then the closer station's signal would arrive first, and the other two at later times, depending upon his distance from them. By measuring this difference in arrival time, the captain or his navigator could translate the information into the ship's position.

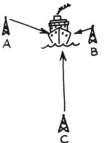

SIGNAL B ARRIVES FIRST.
SIGNAL A ARRIVES SECOND.
SIGNAL C ARRIVES LAST.

There are a number of navigation systems that work in just this way. One such system is Loran-C, which broadcasts signals at 100 kHz. Another is the Omega navigation system, which broadcasts at about 10 kHz. Operating navigation systems at different radio frequencies provides certain advantages. For example, Loran-C can be used for very precise navigation at distances out to about 1600 kilometers from the transmitters, whereas Omega signals can easily cover the whole surface of the earth, but the accuracy of position determination is reduced.

What has time to do with these systems? The answer is that it is crucially important that the radio navigation stations all have clocks that show the same time to a very high accuracy. If they don't, the broadcast signals will not occur at exactly the right instant, and this will cause the ship's navigator to think that he is at one position when he is really at another. Radio waves travel about 300 meters in one microsecond; so if the navigation stations' clocks were off by as little as $\frac{1}{10}$ of 1 ms, the ship's navigator could make an error of many kilometers in plotting the position of his ship.

There is another way that clocks and radio signals can be combined to indicate distance and position. Let's suppose that the captain has on board his ship a clock that is synchronized with a clock at his home port. The home-port clock controls a radio transmission of time signals. The ship's captain will not receive the noon "tick" exactly at noon because of the finite velocity of the radio signal, as we mentioned earlier. Because the captain has a clock synchronized with the home-port clock, he can accurately measure the delay of the signal. If this delay is $\frac{1}{10}$ ms, then he knows that he is about 60 kilometers from the home port.

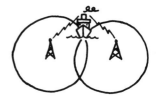

SIGNAL LEAVES SHORE AT THIS TIME AND...        ...ARRIVES AT SHIP 0.1 ms LATER

With two such signals the captain could know that he was at one of two possible points determined by the intersection of two circles, as shown in the sketch. Usually he has a coarse estimate of his position, so he will know which point of intersection is the correct one.

### Navigation by Satellite

We have described a navigation scheme that requires broadcasting signals from three different earth stations. There is no reason, however, that these stations need be on earth; the broadcasts could emanate from three satellites.

Satellites offer various possibilities for navigation systems. An interesting one in actual operation today is the *Transit* satellite navigation system, in which the navigator can determine his position by recording a radio signal from just a single satellite as it passes overhead.

In a sense, receiving a signal from one satellite as it passes by is like measuring signals from many different satellites strung out across the sky, looking at one at a time. The operation of the system depends upon a physical phenomenon called the "Doppler effect," which we have discussed in other connections in this book. We are most familiar with this effect, as we have said, when we hear the whistle of a passing locomotive. As the locomotive approaches, the tone of the whistle is high in pitch, or is "sharp"; then as it passes and moves away, the pitch lowers, or is "flat."

In a similar way, a radio signal from a passing satellite appears high as the satellite approaches, and then lower as the satellite disappears over the horizon. A listener standing at some location other than ours would observe the same phenomenon, but the curve of rise and fall would be different for him. In fact, all observers standing at different locations would record slightly different curves. The position of the satellite is tracked very accurately, so that we could, if time allowed, calculate a catalog of Doppler signals for every point on earth. The navigator, to find his location, would record the rising and falling Doppler signal as the satellite passed overhead, and then find in the catalog the "Doppler curve" that matched his own—and thus could identify his location.

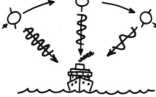

SIGNAL FREQUENCY GOES FROM HIGH TO LOW AS SATELLITE PASSES OVER

Of course, this is not a very practical approach because of the enormous number of calculations that would be required for the entire earth. In practice the user records the Doppler curve as the satellite passes over, and the position of the satellite as broadcast by the satellite itself. Generally the user has at least a vague notion as to his location. He feeds his estimate of his position, along with the satellite location, into a small computer, which calculates the Doppler curve he should be seeing if he is located where he thinks he is.

This calculated curve is compared, by the computer, with the recorded curve. If they are the same, the user has correctly guessed his position. If they are not, the computer makes a new "educated guess" as to his location and repeats the process until there is a good fit between the calculated and the recorded curves. This "best fit" curve is then the curve that yields the best guess as to the user's position.

Scientists in different organizations are working constantly to develop simpler, less expensive methods for keeping track of ships at sea, planes in the air, and even trucks and buses on the highways, through applying time and frequency technology.

## SOME COMMON AND SOME FAR-OUT USES OF TIME AND FREQUENCY TECHNOLOGY

The makers of thermometers, bathroom scales, and tape measures have a fair idea of how many people use their product, who the users are, and what the users do with their measuring devices.

But the suppliers of time and frequency information are like the poet who "shot an arrow into the air." There is no good way of knowing where the "arrows" of their radio broadcasts fall, who picks them up, or whether the "finders" number in the thousands or the tens of thousands. The signal is available everywhere at all times, and remains the same whether ten receivers pick it up or ten thousand. And except for inquiries, complaints, and suggestions for improvements—mostly from users already well known to the broadcasters—those who go to great effort and expense to make their time-measuring metersticks available to the public hear from only a small percentage of those who pick up their wandering "arrows."

They would like to know, however, so that they might make their product more useful and more readily available to more people. So occasionally they make special efforts to survey their "public," and invite users to write to them.

Several such invitations from the National Bureau of Standards have brought many thousands of responses from the usual power company and communication system users; the scientific laboratories, universities, and observatories; the aviation and aerospace industries, and manufacturers and repairers of radio and television equipment and electronics instruments; the watch and clock manufacturers; the military bases. There were scores of responses from private aircraft and yacht and pleasure-boat owners. Many ham radio operators responded, as did a surprising number of persons who classified themselves simply as hobbyists —astronomy or electronics buffs.

Among the specific uses mentioned by these respondents were such things as "moon radar bounce work and satellite tracking," "earth tide measurements," "maintenance of telescope controls and instrumentation," "timing digital clocks," "setting time of day on automatic telephone toll ticketing systems," "synchronizing time-code generators," and "calibrating and synchronizing outdoor time signals in metropolitan areas."

Data processing and correlation, calibration of secondary time and frequency standards, and seismic exploration and data transmission were all very familiar uses. And as more electronic instruments are being developed for use in hospitals and by the medical profession, it was not surprising to have a growing number of users list "biomedical electronics," "instrument time-base calibration for medical monitoring and analyzer equipment," and similar specific medical applications.

Greater use of electronic systems in automobiles has brought automobile mechanics into the ranks of time and frequency technology users. And the proliferation of specialized cameras that take pictures under water, inside organs of the body, or from a thousand kilometers out in space—pictures from microscopic to macroscopic proportions—as well as sophisticated sound recording systems, has greatly increased the need for time and frequency technology among photographic and audiovisual equipment manufacturers and repairers. Oceanography is another rapidly developing science that is finding uses for time and frequency technology.

Time information—both date and time interval—is vitally important, of course, to both manufacturers and repairers of clocks and watches. And with the growing availability and lower prices of fine watches capable of keeping time to a few seconds in a month, more and more jewelers and watch repairmen need time more precise than they can get from the electric clock driven by the power company line, A jeweler on the east coast reported that he telephones the NBS time and frequency information service in Colorado daily to check the watches he is adjusting. The same information, of course, is available via short-wave radio from NBS stations WWV and WWVH; but getting the information by telephone may be simpler and less time consuming, and the signal comes through with little distortion or noise.

Some musicians and organ and musical instrument makers and repairmen reported use of the standard 440 Hz audio tone— the "A" above "middle C" in the musical scale—to check their own secondary standards or to tune their instruments.

Several scientists working on thunderstorm and hailstorm research reported their need for time and frequency information. One specified "coordination of data recordings with time-lapse photography of clouds." Another explained that he used the information to tell him where lightning strikes a power line.

Other responses came from quartz crystal manufacturers, operators and repairers of two-way radio systems, and designers of consumer products—everything from microwave ovens and home entertainment systems to the timers on ranges, cookers, washers, and other home appliances. Even toy manufacturers stated their needs for precise time and frequency technology.

Then there were a few what might be called less serious uses —except to the users, who seemed to be dead serious. An astrologer declared he needed precise time information to render "dependable" charts. Pigeon racers reported using WWV broadcast as a reference point for releasing birds at widely separated locations at the same instant. And persons interested in sports car rallies and precise timing of various kinds of races and other sporting events stated their needs.

Miniaturization and printed circuits have brought a **great** many pieces of inexpensive and useful electronic equipment within easy reach of the average consumer. Electric guitars, radio-controlled garage door closers, "white sound" generators to shut out disturbing noises and soothe one to sleep—who knows what designers and manufacturers will think of next? With these and many other consumer products has come a growing need—even for the man in the street—for better time and frequency technology. This need can only continue to expand, and scientists keep busy working constantly to meet demands.

## B.C.

B.C. BY PERMISSION OF JOHNNY HART AND FIELD ENTERPRISES, INC.

## A NEW DIRECTION

Up to this point we have concentrated on how time is measured, how it is broadcast to almost every point on the surface of the earth, and how it is used in a modern industrial society. In the remainder of this book we shall turn away from the measurement, distribution, and practical uses of time and dip into a number of somewhat unrelated subjects in science and technology where time plays many different roles.

In the next chapter, "Time and Mathematics," for example, we shall explore 17th and 18th Century developments in mathematics, particularly calculus, which provided a new language, uniquely qualified, to describe the motions of objects in concise and powerful ways. Motion, too, is intimately related to time as is indicated by such questions as, what is the period of rotation of the earth, or how long does it take an object to fall a certain distance on the surface of the moon? Then there are always the intriguing and mind-boggling questions concerning the age and evolution of the universe, where the very words, "age" and "evolution," connote time.

These and other subjects discussed in the last chapters of this book are developed so as to cast the spotlight on time, but the reader should be aware that there are many other characters on stage and that time is not the only important player. Modern scientific theories are rich in a variety of concepts and applications, which the reader will quickly discover if he pursues in any depth the subjects which we have examined primarily as they pertain to time (see the suggested reading, page 170).

As a final point, many of these subjects are by no means closed to discussion, but are on the very boundary of research and controversy. In fact, it would appear that every step which leads us toward a better understanding of the nature of time also brings into view new, uncharted territory.

TIME, SCIENCE, AND TECHNOLOGY

# V

# TIME, SCIENCE, AND TECHNOLOGY

# Chapter

| 1 | 2 | 3 | 4 | 5 | 6 | 7 |
| 8 | 9 | 10 | 11 | **12** | 13 | 14 |
| 15 | 16 | 17 | 18 | 19 | 20 | 21 |
| 22 | 23 | 24 | 25 | 26 | 27 | 28 |
| 29 | 30 | 31 | | | | |

## TIME AND MATHEMATICS

It is often said that mathematics is the language of science. But it hasn't always been so. Greek scientists and philosophers were more interested in qualitative discussions of ultimate causes than they were in precise quantitative descriptions of events in nature. The Greeks wanted to know why the stars seemed to circle endlessly around the earth. Aristotle provided an answer—there was an "Unmoved Mover." But Galileo was more interested in *how long* it took a stone to fall a certain distance than he was in *why* the stone fell toward the center of the earth. This change in the thrust of science from the *qualitative why* to the *quantitative how long* or *how much* pushed the need for precise measurement. Along with this shift in emphasis came the development of better instruments for measurement and a new mathematical language to express and interpret the results of these measurements—namely calculus.

One of the most important measurements in science is the measurement of time. Time enters into any formula or equation dealing with objects changing or moving in time. Until the invention of calculus there was no mathematical language expressly tailored to the needs of describing motion and change. We shall describe in some detail in this chapter how the interaction between mathematics—especially calculus—and measurement allows us to construct theories which deeply illuminate the fundamental laws of nature. As we shall see, time and its measurement is a most crucial element in the structure of these theories.

QUALITY
QUANTITY

## TAKING APART AND PUTTING TOGETHER

Man has always been preoccupied with his past and future. An understanding of the past gives him a feeling of identity, and knowledge of the future—insofar as this is possible—helps him chart the most efficient and rewarding course. Much human effort is directed toward trying to see into the future. Whether it's a fortune teller gazing into a crystal ball, a pollster predicting the outcome of an election, or an economist projecting the future of inflation or the stock market, any "expert" on the future has a ready audience.

Science, too, has its own peculiar brand of forecasting. One of the underlying assumptions of a science is that the complex can be understood in terms of a few basic principles, and that the future unfolds from the past according to strict guidelines laid down by these principles. One of the tasks of the scientist, therefore, is to strip his observations down to their bare essentials—to extract the basic principles and to apply them to understanding the past, the present, and the future. The extraction of the principles is usually a "taking apart," or *analysis;* and the application of the principles is a "putting together," or *synthesis.* One of the scientists's most important tools in both of these efforts is mathematics. It helps uncover the well-springs of nature, and having exposed them, helps predict the course of their flow.

## SLICING UP THE PAST AND THE FUTURE—CALCULUS

As we know, in nature everything changes. The stars burn out and our hair grows grey. But as obvious as this fact is, man has difficulty grasping and grappling with change. Change is continuous, and there seems to be no way of pinning down a particular "now."

This struggle is clearly reflected in the development of mathematics. The mathematics of the Greeks was "stuck" in a world of constant shape and length—a world of geometry. The world of numbers continued to be frozen in time until 1666, when Isaac Newton invented the mathematics of change—the *calculus.* With his new tool he was able to extract an "essential quality" of gravitation from Galileo's experimentally discovered law for the distance a rock falls in a given time.

Newton's tool for sifting gravitation from Galileo's formula is called *differentiation,* and differentiation and *integration* are the inverses of each other in the same sense that subtraction is the inverse of addition and division is the inverse of multiplication. Differentiation allows us to pick apart and analyze motion, to discover its instantaneous essence; and integration allows us to synthesize the instantaneous, revealing the full sweep of motion. We might say that differentiation is seeing the trees and integration is seeing the forest.

DIFFERENTIATION=
        TAKING APART

INTEGRATION=
        PUTTING TOGETHER

### Conditions and Rules

Before we get into a more detailed discussion of differential and integral calculus, let's back up a bit and discuss, in general terms, how the mathematical physicist views a problem. Suppose, for example, he wants to analyze the motion of billiard balls. First, he recognizes some obvious facts:

- At any particular moment, all of the balls are moving with certain speeds and directions at particular points on the table.
- The balls are constrained to move within the cushions that bound the table.
- The balls are subject to certain rules that govern their motion—such as, a ball colliding with a cushion at some angle will bounce off at the same angle; or a ball moving in a particular direction will continue in that direction until it strikes a cushion or another ball. The latter is a form of one of Newton's laws of motion, and the former can be *derived* from Newton's laws of motion.

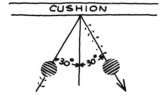

With all of this information, the physicist can predict the motions of the balls. Mathematicians call the statements that characterize the locations and motions of the balls at an instant the

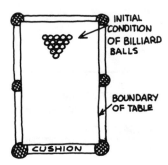

*initial conditions.* And they call the statements that describe the allowable area of motion—in this case the plane of the table bounded by the cushion on four sides—the *boundary conditions.* Obviously the future location of the balls will be very dependent on the shape of the table. A round table will give a result entirely different from a rectangular table.

Thus with a set of initial conditions, boundary conditions, and rules governing the motion, we can predict the locations, speeds, and directions of the balls at any future time. Or we can work backwards to determine these quantities at any earlier time. It's easier if we have a computer!

Having stated that the future and past are related to the "now" of these conditions, how do we proceed? In our example the boundary conditions are simply obtained by measuring the table. Getting the initial conditions is somewhat trickier. From a photograph we determine the locations of the balls at the instant the picture is taken, but we also need to know the *speeds* and *directions* of the balls.

It might occur to us that if we take *two* pictures, one very slightly later than the other—say one-tenth of a second later—we can determine *all* of the initial conditions. From the first photograph we determine the locations of the balls; and from the second, compared with the first, we determine the directions of the balls, as well as their speeds, by measuring the change of position each ball makes in 0.1 second.

Changes in boundary conditions and initial conditions will, of course, alter the future course of events; and it is one of the challenges of physics to deduce from a set of observations which part is due to initial and boundary conditions, and which part is due to the laws governing the process.

Let's apply these ideas now to Galileo's problem of the falling rock. According to legend, Galileo dropped objects from the leaning tower of Pisa and measured their time of fall. But according

B.C. BY PERMISSION OF JOHNNY HART AND FIELD ENTERPRISES, INC.

to his own account, he measured the time it took for bronze balls to roll down a smooth plank—probably because he had no reliable way to measure the relatively short time it takes a rock to fall the length of the tower of Pisa.

In any case, Galileo ultimately came to the conclusion that a free-falling object travels a distance proportional to the time of fall squared, and that the fall time does not depend upon the object's mass. That is, if a rock of any mass falls a certain distance in a particular time interval, it falls four times as far in twice that time interval. More precisely, he discovered that the rock falls a distance "$d$" in meters equal to about 4.9 times the fall time squared, in seconds—or $d = 4.9t^2$.

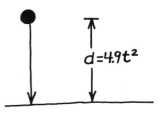

Is this simple formula a law of motion, or is it some combination of laws, boundary conditions, and initial conditions? Let's explore this question.

### Getting at the Truth with Differential Calculus

The method of differential calculus is similar to the scheme we developed for determining the locations, speeds, and directions of the billiard balls at a specified time. Let's look more carefully at the particular problem of determining the speed of a billiard ball.

To get the data we needed, we took two pictures of the moving balls, separated by 0.1 second, and we determined the speed of a ball by measuring the distance it had moved between photos. Let's suppose that a specific ball moved 1 centimeter (cm) —in 0.1 second. We easily see that its speed is 10 cm per second.

1cm IN 0.1 SECOND IS THE SAME AS...

Suppose, however, that we had taken the pictures 0.05 seconds apart. Then between photos the ball moves half the distance, or 0.5 cm. But of course the speed is still 10 cm per second, since 0.5 cm in 0.05 second is the same as 1.0 cm in 0.01 second.

...0.5cm IN 0.05 SECOND

In differential calculus we allow the time between photos to get closer and closer to zero. As we have seen, halving the time between photos does not give us a new speed, because the distance moved by the ball decreases proportionately; so the *ratio* of distance to time between photos always equals 10 cm per second. The essence of differential calculus is to divide one "chunk" by another—in our example, distance divided by time—and allow the two chunks to shrink toward zero. This sounds like very mysterious business, and one might think that the end result of such a process would be dividing nothing by nothing. But this is not the case. Instead, we end up with the rate of motion at a particular instant and point.

This process of letting the chunks shrink toward zero, mathematicians call "taking the limit." By taking the limit we reach the answer we are seeking, namely the magnitude of the motion at a *point* not contaminated by what happened over a *distance*, even though the distance was very small. *Integration* is the reverse process; we take the instantaneous motion and convert it back to distance.

$$\mathcal{d} = 4.9t^2$$

In our particular example, since the billiard ball moves with constant speed, we get the same results with photos that taking the limit produces: 10 cm per second. But in actual fact, the ball won't move with constant speed, but will slow down ever so slightly because of friction. If we wanted to be really accurate in our measurement, we would want our two photos to be separated by the shortest possible time, which approaches the idea of taking the limit.

Let's go back now to Galileo's formula for falling bodies: $d = 4.9t^2$. We would like to reduce this formula to some number that does not change with time—a quantity that characterized the rock's motion independent of initial and boundary conditions. As Galileo's formula stands, it tells us how the *distance fallen* changes with time. Putting a different time in the formula gives a different distance; this formula certainly is not independent of time.

We might think, well, since the distance changes with time, maybe the velocity remains constant. Perhaps the rock falls with the same speed on its journey to the earth.

Differential calculus gives us the answer. It allows us to go from *distance traveled* to *speed of travel*. We can think of this process as putting our formula $d = 4.9t^2$ into a "differentiating machine" which produces a new formula showing how speed, s, changes with time. The process performed by the machine is somewhat akin to the two-photo procedure we discussed in earlier paragraphs. (See also "What's Inside the Differentiating Machine? An Aside on Calculus.") Let's see what happens now:

In goes $(d = 4.9t^2) \rightarrow$ | DIFFERENTIATING MACHINE | $\rightarrow$ out comes $(s = 9.8t)$.

To our disappointment we see that the speed $s$ is not constant; it increases continually with time. For every second the rock falls, the speed *increases* by 9.8 meters per second.

Well, perhaps then the *rate* at which the speed changes, the *acceleration,* "*a,*" is constant. To find out, we run our formula for speed through the differentiating machine:

In goes $(s = 9.8t) \rightarrow$ | DIFFERENTIATING MACHINE | $\rightarrow$ out comes $(a = 9.8)$.

At last we've found a quantity that does not change with time. The acceleration of the rock is always the same. The speed increases at the constant rate of 9.8 meters per second every second. We have hit rock bottom; 9.8 is a number that does not change, and it tells us something about nature because it does not depend upon the height of the tower or the way we dropped the rock.

## WHAT'S INSIDE THE DIFFERENTIATING MACHINE?—AN ASIDE ON CALCULUS

To understand how the differentiating machine works, let's consider a specific example. We'll suppose a rock hits the ground after falling for five seconds, and we would like to know the speed at impact. We start with Galileo's formula that says

$$y = 4.9t^2$$

where $y$ is the distance fallen and $t$ is time. From the formula, we find that the rock has fallen 78.0 meters after 4 seconds and 122 meters after 5 seconds, or that the rock falls 44 meters in its last second before impact. (During this last second, the average speed is 44 meters per second, although at the beginning of the second the rock's true speed is less than this and at the end it is greater.) We repeat this procedure several times, always using Galileo's formula, to obtain the *average* speed during the last ½ second, the last ¼ second, and so forth down to the last 1/16,000 second. The results are shown in the table below.

| TIME INTERVAL (SECONDS) | 1 | $\frac{1}{2}$ | $\frac{1}{4}$ | $\frac{1}{16}$ | $\frac{1}{32}$ | $\frac{1}{64}$ | $\frac{1}{160}$ | $\frac{1}{1600}$ | $\frac{1}{16,000}$ |
|---|---|---|---|---|---|---|---|---|---|
| DISTANCE FALLEN (METERS) | 44 | 23 | 12.00 | 3.03 | 1.52 | 0.76 | 0.30 | 0.03 | 0.003 |
| AVERAGE SPEED (METERS PER SECOND) | 44 | 46 | 47.50 | 48.50 | 48.60 | 48.69 | 48.73 | 48.76 | 48.767 |

If we scan along the bottom row of the table, it appears that the termination speed is 49 meters per second, although with this approach we can never quite prove it. However, with calculus we *can* prove it. Let's see how.

What we want to do is apply calculus to Galileo's formula for distance and turn it into a formula that gives speed for any arbitrary fall time. Since we want a general formula, we shall use symbols, rather than numbers, to derive our new formula.

First, we shall let "$t$" stand for the fall time, and "$\Delta t$" (delta $t$) stand for any short interval of fall time. Thus, we might say that the rock falls for a time t and then falls an additional small interval of time $\Delta t$. Similarly, we shall let $\Delta y$ stand for the distance the rock falls during the short interval of time $\Delta t$.

Next, using Galileo's formula, we want to find a formula for $\Delta y$, the distance the rock falls in $\Delta t$ seconds.

We start with $y = 4.9t^2$.

Then it follows that $y + \Delta y = 4.9\,(t + \Delta t)^2 =$
$$4.9t^2 + 9.8t\Delta t + 4.9\,(\Delta t)^2.$$

Subtracting the first formula for $y$ from the second formula for $y + \Delta y$ we see that

$$\Delta y = 9.8t\,\Delta t + 4.9\,(\Delta t)^2.$$

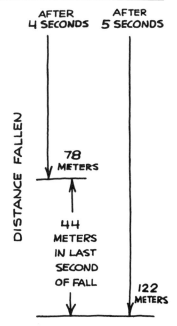

AFTER 4 SECONDS    AFTER 5 SECONDS

DISTANCE FALLEN

78 METERS

44 METERS IN LAST SECOND OF FALL

122 METERS

Since the distance $\Delta y$ is covered in a time $\Delta t$, the average speed over the distance is $\Delta y/\Delta t$.

$$\Delta y/\Delta t = \frac{9.8t\,\Delta t + 4.9\,(\Delta t)^2}{\Delta t} = 9.8t + 4.9\,\Delta t.$$

Finally, what we want to know is not the average speed over the distance $\Delta y$, but the instantaneous change of distance with respect to time or, equivalently, the speed at a point. We do this by letting $\Delta t$ go to zero; or as the mathematicians say, we "take the limit" as $\Delta t$ approaches zero, which is

instantaneous speed $= \underset{\Delta t \to 0}{\text{limit}}\,(9.8\,t + 4.9\,\Delta t) = 9.8t$ or speed, s $= 9.8t$.

Let's see how this new formula works. We put 5 seconds into our formula and we obtain $s = 9.8 \times 5$, $= 49$ meters per second, which is precisely what we anticipated from our table.

The instantaneous change of distance, $y$, with respect to time, $t$, is called the "derivative" of $y$ with respect to $t$. It is written $dy/dt$, which is just in our particular example, a shorthand notation for the process, $\underset{\Delta t \to 0}{\text{limit}}$.

In compact mathematical symbols, the derivative with respect to time, $t$, of the distance fallen, $y$ ($= 4.9t^2$), is expressed as:

$$\frac{d(y)}{dt} = \frac{d(4.9t^2)}{dt} = 9.8t = \text{speed} = \text{s.}$$

Or as we said on page 118:

in goes $(d = 4.9t^2) \rightarrow \boxed{\begin{array}{c}\textbf{DIFFERENTIATING}\\ \textbf{MACHINE}\end{array}} \rightarrow$ out comes $(s = 9.8t)$.

That part of calculus which is devoted to taking derivatives is called "differential calculus," and the inverse of differential calculus is called "integral calculus." If we had an "integrating" machine for integral calculus, it would take the formula for speed and turn it back into the formula for distance:

In goes $(s = 9.8t) \rightarrow \boxed{\begin{array}{c}\textbf{INTEGRATING}\\ \textbf{MACHINE}\end{array}} \rightarrow$ out comes $(y = 4.9t^2)$.

We shall not go into the details here, but the process of going from $(s = 9.8t)$ .to $(y = 4.9t^2)$—that is, integrating speed with respect to time——is somewhat similar to the exercise we have just completed. The difference is that instead of using Galileo's formula to compute the average speed over each small interval of time, we use our new formula for speed $(s = 9.8t)$ to determine the distance fallen over many short consecutive intervals of time: then we add up all of these intervals to obtain the total distance fallen, and finally, let the intervals of time go to zero (take the limit) to obtain the exact result. To complete the integration process correctly for

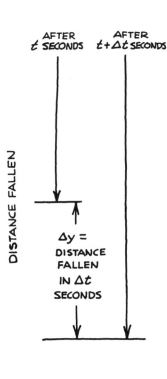

AFTER
$t$ SECONDS

AFTER
$t + \Delta t$ SECONDS

$\Delta y =$
DISTANCE
FALLEN
IN $\Delta t$
SECONDS

DISTANCE FALLEN

a particular problem, we must know the initial and boundary conditions. For example, we can compute correctly the speed of a rock after it has fallen for 10 seconds from the formula $y = 4.9t^2$ only if the rock starts its fall from rest. If it has an initial downward motion—such as would result if we threw the rock downward instead of simply releasing it from our hand—then we must include this fact in our calculation if we are to obtain the correct answer. That is, we must know the initial condition that the rock left our hand at, say, 14 kilometers per hour as well as the formula $y = 4.9t^2$ to obtain the correct answer.

## NEWTON'S LAW OF GRAVITATION

Of course, nature might have been different; we might have found that the acceleration increased with time and that the rate of increase of acceleration with time was constant. But this is not the case. After two "peelings," Newton had discovered that *gravitational pull* produces constant acceleration independent of time.

## THE WIZARD OF ID

THE WIZARD OF ID BY PERMISSION OF
JOHNNY HART AND FIELD ENTERPRISES, INC.

Armed with calculus and Galileo's and others' measurements, he was able to develop his famous *law of gravitation*. He demonstrated that this law applied not only to falling rocks, but also to the solar system and the stars. The latter step required the application of *integral* calculus. By working the whole process backward—*integrating* the instantaneous motions of the planets—he proved that they had to move around the sun in ellipses. By looking through the "microscope" of differentiation, Newton was able to discover the essence of falling bodies; and by looking through the "telescope" of integration, he was able to see the planets circling the sun.

For whatever reasons, Newton kept his invention of calculus to himself, and it was invented again some ten years later by Gottfried Wilhelm von Leibnitz, a German mathematician. Even then, Newton did not publish his version for another 20 years. Leibnitz's symbolism was easier to manage than Newton's, so calculus developed at a faster rate on the Continent than it did in England. In fact, a rivalry developed between the two groups, with

A

BEAD →

WIRE →

B

SHORTEST TIME
(BRACHISTOCHRONE)
PROBLEM

each group trying to stump the other by posing difficult mathematical questions.

One problem posed by the Continental side concerned the shape a wire should have (not straight up and down) so that a bead sliding down it would reach the bottom in the shortest possible time. Newton spent an evening working out the solution, which was relayed anonymously to the Continent. One of Leibnitz's colleagues, who had posed the problem, received the solution and reportedly said, "I recognize the lion by his paw."

Although Newton's laws of motion and gravitation can be summed up in a few simple mathematical statements, it took many great applied mathematicians—men such as Leonhard Euler, Louis Lagrange, and William Hamilton—another 150 years to work out the full consequences of Newton's ideas. Rich as Newton's work was, even he realized that there was much to be done in other fields, especially electricity, magnetism, and light. It was some two hundred years after Newton before substantial progress was made in these areas.

Of course even later, Newton's laws were overturned by Einstein's *relativity;* and most recently quantum mechanics, with its rules governing the microscopic world, has come into play. Each new understanding of nature has led to dramatic gains in the search for perfect *time.*

# Chapter

| 1 | 2 | 3 | 4 | 5 | 6 | 7 |
|---|---|---|---|---|---|---|
| 8 | 9 | 10 | 11 | 12 | **13** | 14 |
| 15 | 16 | 17 | 18 | 19 | 20 | 21 |
| 22 | 23 | 24 | 25 | 26 | 27 | 28 |
| 29 | 30 | 31 | | | | |

# TIME AND PHYSICS

Although much of the early interest in time sprang from religious activities, the "high priests" of time had worked out remarkable schemes for predicting various astronomical events such as the summer solstice, the winter solstice, and motions of the constellations in the sky throughout the year. In later times, with the development of commerce and naval activities on the seas and the need for improved navigation methods, the interest in time turned somewhat from the religious to the secular. But it makes little difference what the application, religious or secular, the tools for "unraveling" the fabric of time are the same.

As we have seen, time measurement has been intimately connected with astronomy for many centuries, and within our own lifetime we have begun to look at the atom in connection with the measurement of time. It is rather curious that we have made a quantum jump from one end of the cosmological scale to the other —from stars to atoms—in our search for the perfect clock. The motions of pendulums, planets, and stars are understood in terms of Newton's laws of motion and gravity—or with even greater refinement in terms of Einstein's General Theory of Relativity; and the world of the atom is understood in terms of the principles of quantum mechanics. As yet, however, no one has been able to come up with one all-encompassing set of laws that will explain man's observations of nature from the smallest to the largest— from the atoms to Andromeda. Perhaps the ultimate goal of science is to achieve this unified view. Or in the words of the artist-poet William Blake, the task of science is

*To see a world in a grain of sand*
*And a heaven in a wild flower,*
*Hold infinity in the palm of your hand*
*And eternity in an hour.*

Auguries of Innocence

BETTER TIME
MEASUREMENT
IMPROVES
UNDERSTANDING
OF NATURE

BETTER UNDERSTANDING
OF NATURE
IMPROVES TIME
MEASUREMENT

Until this goal is reached, the science of physics will continue to ask many difficult questions about the big and the small, and short and the long, the past and the future.

As we saw in our discussion of man-made clocks, the developments of scientific knowledge and of time measurement have moved forward hand in hand; more accurate time measurements have led to a better understanding of nature, which in turn has led to better instruments for measuring time. Improvements in the measurement of time have made it possible for the science of physics to expand its horizons enormously; and because of this, modern physics has several very important things to tell us about time:

- Time is relative, not absolute.
- Time has direction.
- Our measurement of time is limited in a very fundamental way by the laws of nature.
- A time scale based upon one particular set of laws from physics is not necessarily the same time scale that would be generated by another set of laws from physics.

## TIME IS RELATIVE

Isaac Newton stated that time and space are absolute. By this he meant that the laws of motion are such that all events in nature appear to proceed in the same fashion and order no matter what the observer's location and motion. This means that all clocks synchronized with each other should constantly show the same time.

Albert Einstein came to the conclusion that Newton was wrong. Certain peculiar motions in the movement of the planet Mercury around the sun could not be explained according to Newton's ideas. By assuming that space and time are *not* absolute, Einstein was able to formulate new laws of motion that explained the observed orbit of Mercury.

But what have Einstein's ideas to do with clocks not all showing the same time? First Einstein stated that if two spaceships approach each other, meet, and pass in outer space, there is no experiment that can be performed to determine which spaceship is moving and which is standing still. Each ship's captain can assert that his ship is standing still and the other's is moving. Neither captain can prove the other wrong. We've all had the experience of being in a train or other vehicle that is standing still, but feeling sure it is in motion when another vehicle close beside us moves past. Only when we look around for some other, stationary, object to *relate* to can we be sure that it is the vehicle beside us, and not our own, that is in motion.

We must emphasize here that we are not attempting to prove Einstein's idea true; we are merely stating what he had to assume in order to explain what happens in nature. Let's suppose now that each of our two spaceships is equipped with a clock. It's a rather special kind of clock that consists simply of two mirrors, facing each other and separated by a distance of 5 centimeters (cm). The period of the clock is determined by a pulse of light that is simply bouncing back and forth between the two surfaces of the mirrors. Light travels about 30 cm in $10^{-9}$ seconds (one nanosecond); so a round trip takes $\frac{1}{3}$ nanoseconds.

The captain traveling on ship number one would say that the clock on ship number two is "ticking" more slowly than his because its pulse of light traveled more than 10 cm in its round trip. But the captain in ship number two would be equally correct in making the same statement about the clock on ship number one. Each captain views the other's clock *relative* to his. And since, according to Einstein's statement, there is no way to tell which ship is moving and which is standing still, one captain's conclusion is as valid as the other's. Thus we see that time is relative—that is, the time we see depends on our point of view.

To explore this idea a little further, let's consider the extreme case of two spaceships meeting and passing each other at the speed of light. What will each captain say about the other's clock? We could use the mathematics of relativity to solve this problem, but we can also go straight to the answer in a fairly easy way:

When we look at a clock on the wall, what we are really seeing is the light reflected from the face of the clock. Let's suppose that the clock shows noon, and that at that moment we move away from the clock at the speed of light. We will then be moving along with the *light image* of the noon face of the clock, and any later time shown on the clock will be carried by a light image that is also moving at the speed of light. But that image will never catch up with us; all we will ever see is the noon face. In other words, for us time will be frozen, at a standstill.

There are other interesting implications in this concept of time as relative to one's own location and movement. The fact that each captain of our two spaceships sees the other's clock "ticking" more slowly than his own is explained by Einstein's "Special Theory of Relativity," which is concerned with *uniform* relative motion between objects. Sometime later, Einstein developed his "General Theory of Relativity," which took gravitation into account. In this case he found that the ticking rate of a clock is influenced by gravitation. He predicted that a clock in a strong gravitational potential, as is the case near the sun, would appear to us to run slow.

To see why this is so, we go back to our two rocket ships. This time, let's suppose that one of the ships is stopped a certain distance from the sun. At this point there will be a gravitational field "seen" by the spaceship and its contents, including the mirror clock. Let's suppose the other spaceship is falling freely in space toward the sun. Objects in this spaceship will float around freely

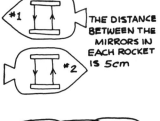

THE DISTANCE BETWEEN THE MIRRORS IN EACH ROCKET IS 5cm

AS THE TWO ROCKET SHIPS PASS EACH OTHER, THE CAPTAIN OF #1 SAYS THAT THE LIGHT PULSE IN #2 TRAVELS MORE THAN 10 cm IN ITS BACK AND FORTH JOURNEY

in the cabin as though there were zero gravitational field—just as we have seen from televised shots of the astronauts on their way to the moon.

Let's suppose now that the falling spaceship passes the stationary spaceship on its journey toward the sun, in such a way that the captain in the free-falling spaceship can see the clock in the other spaceship. Because of the relative motion, he will **again** say that the other clock is running more slowly than his own. And he might go on to explain this observation by concluding that the clock where there is a gravitational field runs more slowly **than** one where there is zero gravitational field.

We can use these observations about clocks from Special Relativity and General Relativity theories to obtain an interesting result. Suppose we put a clock in a satellite. The higher the satellite is above the surface of the earth, the faster the clock will run because of the reduced gravitational potential of the earth. Furthermore, there will also be a change in the rate of the clock caused by the *relative motion* of the satellite and the earth. The difference in relative motion increases as the height of the satellite increases. Thus these two relativistic effects are working against each other. At a height of about 3300 kilometers above the surface of the earth, the two effects cancel each other. So a clock there would run at the same rate as a clock on the surface of the earth.

## TIME HAS DIRECTION

If we were to make a movie of two billiard balls bouncing back and forth on a billiard table, and then show the movie backwards, we would not notice anything strange. We would see the two balls approaching each other, heading toward the edges of the table, bouncing off, passing each other, and so forth. No laws of motion would appear to have been violated. But if we make a movie of an egg falling and smashing against the floor, and then show *this* film backwards, something is clearly wrong. Smashed eggs do not in our experience come back together to make a perfect egg and then float up to someone's hand.

In the egg movie there is very much a sense of *time direction,* whereas in the billiard ball movie there is not. It would appear that the sense of time direction is somehow related to the probability or improbability of events. If we film the billiard balls over a longer period of time, for example, so that we see the balls slowing down and finally coming to rest—and then show the movie backwards, we would realize that it was running backwards. Billiard balls don't start moving from rest and gradually increase their motion—at least not with any great probability.

Again the direction of time is determined by the probable sequence of events. The reason the balls slow down is that friction between the balls and the table gradually converts the ordered rolling energy of the balls into the heating up of the table and balls—ever so slightly. Or more precisely, the ordered motion of the balls is going into disordered motion. A measure of this disordered motion is called *entropy*. Entropy involves *time* and the fact that time "moves" in only one direction.

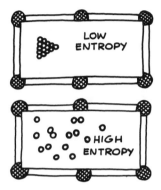

Systems that are highly organized have low entropy, and vice versa. To consider our billiard table further, let's suppose we start by racking up the balls into the familiar triangular shape. At this point the balls are highly organized and remain so until we "break" them. Even after the break, we can perceive a certain organization, but after a few plays the organization has disappeared into the random arrangement of the balls. The entropy of the balls has gone from low to high.

Now let's suppose that we had filmed this sequence from the initial break until well after disorder had set in. Then we run the film backwards. During the first part of the showing, we are watching the balls' motion during a period when they are completely randomized, and in this interval we cannot tell the difference between showing the film forward or backward. In the physicist's jargon, after the entropy of a system is maximized we cannot detect any direction of time flow.

As we continue to watch the film, however, we approach the moment when the balls were highly organized into a compact triangle. And as we come nearer to this moment we can certainly detect a difference between the film running forward or backward. Now we can assign direction to "time's arrow."

There is another point we can bring out by comparing this observation with our earlier discussion of two balls. We noticed that with two balls, we were not able to detect a time direction, but with many balls we can assign a meaning to time direction. With just *two* balls we are not surprised when they collide with each other and move off, but when *many* balls are involved it is highly improbable that they will eventually regroup to form a compact triangle—unless we are running the film backward.

## TIME MEASUREMENT IS LIMITED

We have discussed why it was necessary for Einstein to modify Newton's laws of motion. Some years later scientists discovered that it was necessary to modify Newton's laws to explain observations of objects at the other end of the size scale from planets and stars—namely, atoms. But the modifications were different from those that Einstein made.

One of the implications of these modifications is that there is a limit to how precisely time can be measured under certain conditions. The implication is that the more we want to know about *what* happened, the less we can know *when* it happened, and vice versa. It's a kind of "you can't have your cake and eat it" type of law.

We can get a feeling for it by considering the following problem: Let's suppose we would like to know the exact instant when a BB, shot from an air rifle, passes a certain point in space. As the BB passes this point, we could have it trigger a fine hair mechanism that sets off a high-speed flash photo. Behind the BB is a wall clock whose picture is taken along with the BB's picture. In the high-speed photo we would see the BB suspended in motion, and the wall clock would indicate the time at the moment the picture was taken.

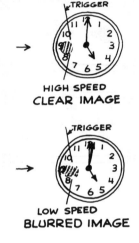

**HIGH SPEED**
**CLEAR IMAGE**

**LOW SPEED**
**BLURRED IMAGE**

But let's suppose we wanted to know something about the *direction* the BB is moving, but we were still limited to one picture. We could take a slower picture, which would show a blurred image of the BB, and from this image we could determine the direction of movement. But now the second hand on the clock would be blurred also, and we could not know the exact *time* when the BB crossed the mark.

Both the time when an event happens and the duration of time it occupies can be measured quite precisely. But the greater the degree of precision, the less other information can be gathered. This fact, which scientists call the "uncertainty principle," seems to be a fundamental feature of nature.

From quantum mechanics a more precise expression of the uncertainty principle is that the more we know about the "energy" of a process, the less we know about when it happened, and vice versa. We can apply this statement directly to an atom emitting a photon of radiation, such as a hydrogen atom in a hydrogen maser. According to the uncertainty principle, the more precisely we know the amount of energy emitted by the atom, the less we know about when it happened.

In Chapter 5 we found that the frequency of the radiated energy is related in a precise way to the "quantum" of energy emitted: The bigger the quantum of energy, the higher the frequency emitted. But now we see that there is another application of this frequency-energy relationship. If we know the magnitude of the quantum of energy quite accurately, then we know the frequency radiated quite accurately, and vice versa.

But the uncertainty principle tells us that to know the energy precisely—and thus the frequency precisely—means that we won't know very much about *when* the emission took place. The situation is somewhat like water flowing out of a reservoir. If the water runs out very slowly, we can measure its rate of flow accurately;

but this flow will be spread over a long period of time, so we have no distinct notion of the process having a precise beginning and ending. But if the dam is blown up, a huge crest of water will surge downstream; and as the crest passes by, we will have no doubt that something happened and when it happened—but we will have great difficulty in measuring the rate at which the water flows by.

In the case of the atom, the energy leaking away slowly means that we can measure its frequency precisely. We have already observed a similar idea in Chapter 5, in our discussion of the cesium-beam tube resonator. We said that the longer the time the atom spends drifting down the beam tube, the more precisely we could determine its resonance frequency; or alternatively, the longer the time the atom spends in the beam tube, the higher the "$Q$" of the resonator. Thus, both from our discussion of resonators in Chapter 5, and also from quantum mechanics, we come to a conclusion that makes sense: The longer we observe a resonator, the better we know its frequency.

As a final comment, we can relate our discussion here to the spontaneous emission of an atom, which we also discussed in Chapter 5. Atoms, we observed, have a "natural" lifetime. That is, left alone they will eventually spontaneously emit a photon of radiation; but this lifetime varies from atom to atom and also depends upon the particular energy state of the atom. If an atom has a very short natural lifetime, we will be *less* uncertain about when it will emit energy than if it had a very long lifetime. Invoking the uncertainty principle, atoms with short lifetimes emit uncertain amounts of energy, and thus the frequency is uncertain; and atoms with long lifetimes emit packets of energy whose values are well known, and thus the frequency is well established.

We see in a sense, then, that each atom has its own natural $Q$. Atoms with long natural lifetimes correspond to pendulums with long decay times, and thus high $Q$'s; and atoms with short natural lifetimes correspond to pendulums with short decay times, and thus low $Q$'s.

We should emphasize, however, that although a particular energy transition of an atom may correspond to a relatively low $Q$ as compared to other transitions of other atoms, this of itself is not necessarily a serious obstacle to clock building. For atomic resonators contain many millions of atoms, and what we observe is an "average" result, which smooths out the fluctuations associated with emissions from particular atoms. The only limitation would appear to be the one that we have already pointed out in Chapter 5, in our discussion of the limitations of atomic resonators: As we go to higher and higher atomic resonant frequencies, the natural lifetime of the atom—or decay time—may be so short that it would be difficult to observe the atom before it spontaneously decayed.

**FROM THIS
TO THIS**

## ATOMIC AND GRAVITATIONAL CLOCKS

We have seen that there is no single theory in science that explains both the macroscopic world of heavenly bodies and the microscopic world of the atom. Gravity governs the motions of the stars, galaxies, and pendulums; atoms come under the jurisdiction of quantum mechanics.

In our history of the development of clocks, we have seen that there has been a dramatic change in the last decade or so. We have gone from clocks whose resonators were based on swinging pendulums—or other mechanical devices—to resonators based on atomic phenomena. We have gone from the macro-world to the micro-world with a concomitant change in the laws that govern the clocks' resonators.

This diversity raises an interesting question: Do clocks understood in terms of Newton's law of motion and gravitation keep the same time as those based on quantum theory? The atomic second, we recall, was defined as nearly equal to the ephemeris—or "gravity"—second for the year 1900. But will this relationship be true a million years from now—or even a thousand? Might not the atomic second and the gravitational second slowly drift apart?

The answer to this challenging question is embedded in the deeper question of the relationship between the laws describing the macro-world and those describing the micro-world. In the laws of both there are numbers called *physical constants*, which are assumed not to change in time. One such constant is the velocity of light; another is the gravitational constant "$G$." $G$ appears in Newton's law of gravitation, according to which the gravitational attraction between two objects is proportional to the product of their masses and inversely proportional to the square of the distance between them. Thus if we write $M_1$ and $M_2$ for the two masses, and the distance between them $D$, Newton's law reads

$$\text{FORCE} = F = G\frac{M_1 \, M_2}{D^2}$$

$$F = \frac{GM_1 \, M_2}{D^2}$$

To get the correct answer for the force, $F$, we have to introduce $G$; and $G$ is a number that we must determine experimentally. There is no scientific theory that allows us to calculate $G$.

We have a similar situation in quantum mechanics. We have already learned that energy, $E$, is related to frequency, $f$; mathematically the expression is $E = hf$, where $h$ is another constant—Plank's constant—which must be determined experimentally. If in some unknown way, $G$ or $h$ is changing with time, then time kept by gravitational clocks and atomic clocks will diverge. For if $G$ is changing slowly with time, a pendulum clock under the influence of gravitation will slowly change in period. Similarly, a changing $h$ will cause the period of an atomic clock to drift. At present there is no experimental evidence that this is happening, but the problem is being actively investigated.

If $G$ and $h$ diverge in just the "right" way, we could get some rather strange results. Let's suppose gravitational time is slowly decreasing with respect to atomic time. We can't really say which scale is "correct"; one is just as "true" as the other. But let's take the atomic time scale as our reference scale and see how the gravitational scale changes with respect to it.

We'll assume that the rate of the gravitational clock doubles every thousand million years—every billion years—with respect to the atomic clock. To keep our example as simple as possible, let's assume that the rate of the gravitational clock does not change smoothly but occurs in jumps at the end of each billion years. Thus, one billion years ago the pendulum or gravitational clock was running only half as fast as the atomic clock. As we keep moving into the past, in billion-year intervals, we could tabulate the total time as measured by the two kinds of clocks as follows:

Accumulated Atomic Clock Time = 1 billion years + 1 billion years + 1 billion years + and so forth indefinitely.

Accumulated Pendulum Clock Time = 1 billion years + ½ billion years + ¼ billion years + ⅛ billion years + and so forth indefinitely.

As we go further and further into the past, the accumulated atomic time approaches infinity, but the accumulated pendulum clock time does not; it approaches 2 billion years:

$$1 + \frac{1}{2} + \frac{1}{4} + \frac{1}{8} + \frac{1}{16} + \frac{1}{32} + \frac{1}{64} \ldots = 2.$$

The arithmetic is similar to the problem we discussed earlier, on page 117, where we saw that the speed of a rock hitting the ground approached 160 feet per second, as we kept computing its average speed at intervals successively closer to the ground. Thus, in our example, gravitational time points to a definite *origin* of time, and atomic time does not.

The example we have chosen is, of course, just one of many possibilities, and we picked it for its dramatic qualities. But it does illustrate that questions relating to the measurement of time must be carefully considered. To ask questions about time and not specify how we will measure it is most probably an empty exercise.

### THE DIRECTION OF TIME AND SYMMETRIES IN NATURE—AN ASIDE

We can relate this discussion of the billiard balls and time's direction to what was said about mathematics and time on page 115. We recall that a mathematician-scientist characterizes a problem by the initial conditions, the boundary conditions, and the laws that govern the process he is investigating. The direction of time's arrow is a consequence of initial conditions, and not a consequence of the laws governing the motions of the balls. It is the initial condition that all of the balls start from a triangular nest and proceed toward random positions over the surface of the table that gives a sense of time's direction.

If the balls had started out randomly—that is if the initial condition had been a random placement of the balls, with random speeds and directions of motion—then we would have no perception of time's direction. But the *laws* governing the motion are the same whether the balls are initially grouped or scattered over the surface of the table.

These observations bring up a very interesting question: Is there no sense of time direction in a random universe? From the preceding discussion, it would appear that there is not. But we cannot give a final answer to this question. Until 1964, it appeared that there was no law in nature that had any sense of time direction, and that time's arrow is simply a consequence of the fact that nature is moving from order to disorder. That is, in the distant past the universe was compact and ordered, and we are now some 10 to 20 billion years down the path to disorder. We shall discuss this again later, in connection with the "big bang" theory of the universe.

## THE STRUGGLE TO PRESERVE SYMMETRY

To dig deeper into this question of time's direction, we move into an area of physics that is on the frontier of experimentation and theoretical development. This means the subject is highly controversial and by no means resolved. All we can do at this point is attempt to describe the present state of affairs, and the reader's guess as to what the future will bring to light is perhaps as good as anyone's.

As we have said, there was no evidence until 1964 that the laws of nature contained the least indication of the direction of time. But ten years before that, in 1954, two physicists, T. D. Lee and C. N. Yang, of the Institute for Advanced Study, at Princeton, inadvertently opened up new speculation on this subject.

A very powerful notion in physics is that there is a certain symmetry inherent in nature and that the detection of symmetries greatly clarifies our understanding of nature. Let's return to our two billiard balls for an example. We found that there was no way to determine the direction of time by running the film forward and backward—assuming that the table is frictionless. The laws governing the interactions and motions of the balls are not sensitive to time's direction. We could extend this test to any of the laws of

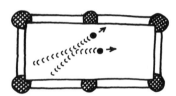

THIS IS A MOVIE OF TWO BALLS COLLIDING

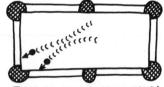

THIS IS THAT MOVIE SHOWN BACKWARD

BOTH THE MOVIE AND ITS BACKWARD VERSION CAN OCCUR IN NATURE

TIME INVARIANCE

nature that we wished to investigate by making movies of processes governed by the law under consideration. As long as we could not tell the difference between running the film backward and forward, we could say that the laws were insensitive to time's direction, or in the language of physics, time "invariance"—or "*T*" invariance—is preserved.

But "*T*" invariance is not the only kind of symmetry in nature. Another kind of symmetry is what we might call left-hand, right-hand symmetry. We can test for this kind of symmetry by performing two experiments. First, we set up the apparatus to perform a certain experiment, and we observe the result of this experiment. Then we set up our apparatus as it would appear in the mirror image of our first experiment and observe the result of *this* experiment. If left-right symmetry is preserved, then the result of our second experiment will be just what we observe by watching the result of our first experiment in a mirror. If such a result is obtained, then we know that left-right symmetry is preserved—or as the physicist would say, there is *parity* between left and right, or "*P*" invariance is preserved.

BOTH THE MOTION OF THE BALL AND ITS MIRROR IMAGE ARE ALLOWED IN NATURE

LEFT-RIGHT INVARIANCE OR "P" INVARIANCE

Until 1956, everyone believed that left-right parity was always preserved. No experiment gave any evidence to the contrary. But Lee and Yang, in order to explain a phenomenon that was puzzling scientists studying certain tiny atomic particles, proposed that parity is not *always* preserved.

The issue was settled by an experiment performed in 1957 at the National Bureau of Standards facilities by Madame Chien-Shuing Wu, Ernest Ambler, R. W. Hayward, D. D. Hoppes, and R. P. Hudson—scientists from Columbia University—and the NBS in Washington, D. C. We shall not go into the details here, but the result conclusively proved that parity had fallen; a pivotal concept of quantum mechanics—and even the common-sense world—had been destroyed.

The violation of parity—"*P*" invariance—greatly disturbed physicists, and they looked for a way out. To glimpse the path they took we must consider a third kind of symmetry principle in addition to "*T*" and "*P*" invariance. This is called *charge conjugation,* or "*C*" invariance. In nature there exists, for every kind of particle, an opposite number called an *antiparticle*. These antiparticles have the same properties as their counterparts except that their electric charges, if any, are opposite in sign. For example, the antiparticle for the *electron*, which has a negative charge, is the *positron*, which has a positive charge. When a particle encounters its antiparticle they can both disappear into a flash of electromagnetic energy, according to Einstein's famous equation $E = mc^2$. Antiparticles were predicted theoretically by the English physicist Paul Dirac, in 1928, and were detected experimentally in 1932 by American physicist Carl Anderson, who was studying cosmic rays.

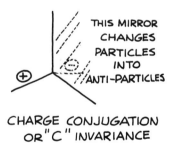

THIS MIRROR CHANGES PARTICLES INTO ANTI-PARTICLES

CHARGE CONJUGATION OR "C" INVARIANCE

But what has all of this to do with the violation of parity—and of even more concern to us—*time reversal?* Let's deal with the parity question first. As we have said, experiments were performed that demonstrated that parity was violated, which was a very disturbing result. But physicists were able to salvage symmetry in a very ingenious way: If the mirror is replaced by a new kind of mirror, which not only gives a mirror image but also transforms the particles in the experiment into their corresponding antiparticles, the symmetry is preserved. In other words, we obtain a result that does not violate a new kind of symmetry. We could say that nature's mirror not only changes right to left, but also reverses matter into antimatter. Thus charge invariance and parity invariance taken together are preserved—"CP" invariance is preserved—and physicists were much relieved.

This inner peace was short lived, however. In 1964 J. H. Christenson and his colleagues at Princeton University performed an experiment that gave a result that could not be accounted for even by a mirror that replaced matter by antimatter, and $CP$ symmetry was broken. But this turns out to have an implication for time invariance, or "$T$" invariance. From relativity and quantum mechanics, cornerstones of modern physics, comes a super-symmetry principle that says: If we have a mirror that changes left to right, exchanges matter for antimatter, and on top of all this, causes time to run backward in the sense of showing a movie backward—then we get a result that should be allowed in nature.

## SYMMETRIES
### TIME INVARIANCE
### LEFT RIGHT
### CHARGE CONJUGATION

We must underline here that this super-symmetry principle is not based on a few controversial experiments undertaken in the murky corners of physics, but is a clear implication of relativity and quantum mechanics; and to deny this super principle would be to undermine the whole of modern physics. The implication of the experiment in 1964 that demonstrated a violation of "CP" invariance requires that *time invariance symmetry—"T"* symmetry—be broken, if the super principle is to be preserved. Nobody to date has actually observed "$T$" invariance symmetry violated. It is only inferred from the broken $CP$ symmetry experiment combined with the super $CPT$ principle.

It is not clear what all of these broken symmetries mean for man in his everyday life. But such difficulties in the past have always created a challenge leading to new and unexpected insight into the underlying processes of nature. The violations of the symmetry principles we have discussed represent only a very few exceptions to the results generally obtained, but these minute discrepancies have many times led to revolutionary new ways of looking at nature.

# TIME AND ASTRONOMY

We have seen that the measurement and determination of time are inseparably related to astronomy. Another facet of this relationship, which sheds light on the evolution of the universe and the objects it contains, has been revealed over the past few decades. In this chapter we shall see how theory combined with observations has allowed us to estimate the age of the universe. We shall discuss some "stars" that transmit signals like "clockwork," and we shall discuss a peculiar kind of star to which the full force of relativity theory must be applied if we are to understand the flow of time in the vicinity of such a star. And finally we shall discuss a new technique of radio astronomy that became possible only with the development of atomic clocks, and that has interesting applications outside of radio astronomy.

## MEASURING THE AGE OF THE UNIVERSE

In 1648 Irish Archbishop Usher asserted that the universe was formed on Sunday, October 23, 4004 B. C. Since then there have been numerous estimates of the age of the universe, and each new figure places the origin back in a more distant time. In the 19th Century, Lord Kelvin estimated that it had taken the earth 20 to 40 million years to cool from its initial temperature to its present temperature. In the 1930's, radioactive dating of rocks settled on two billion years, and the most recent estimates for the age of the universe lie between 10 and 20 billion years.

These newest estimates are developed along two lines of thought and observation: The first relates the age of the universe to the speeds, away from the earth, of distant galaxies. The second

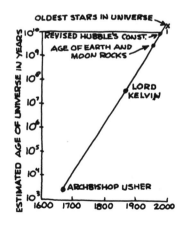

is obtained from observations of the makeup of the universe that peg it as being at a particular point in *time* along its evolutionary "track."

### The Expanding Universe—Time Equals Distance

Throughout much of history, man has tended to think of the universe as enduring "from everlasting to everlasting." But in 1915 Einstein applied his theory of general relativity to the problem of the evolution of the universe and reluctantly came to the conclusion that the universe is dynamic and expanding. In fact, he was so dubious about his conclusion that he introduced a new term into his equations—the "cosmological term"—to prevent his equations from predicting this expansion. Then in 1929, some 14 years later, the American astronomer Edwin Hubble discovered that the universe was indeed expanding, and Einstein is reported to have said that the cosmological term was "the biggest blunder of my life."

We have already encountered the technique used by Hubble to discover the expanding universe. It is based on the Doppler effect, whereby the whistle of an approaching train seems to have its frequency shifted upward, and then shifted downward as the train moves away. Hubble was investigating the light from a number of celestial objects when he noticed that certain aspects of the light spectrum were shifted to lower frequencies, as though the radiating objects were moving away from the earth at high speeds. Furthermore, the more distant the object, the greater its speed away from the earth.

With Hubble's discovery of the relationship between distance and speed, it was possible to estimate an age of the universe. The fact that all objects were moving away from the earth meant that all celestial objects had at some time in the distant past originated from one point. The observed distance to the objects, with their corresponding recessional speeds, when extrapolated backward in time, indicated an origin about 20 billion years ago. Of course, we might suspect that the recessional speeds have been slowing down with time; so the age of the universe could be less than that derived from the presently measured speeds. In fact, using the evolutionary line of reasoning to estimate the age of the universe, we find that this seems to be the case.

### Big Bang or Steady State?

Scientists have developed theories for the evolution of the universe. And according to these theories, the universe evolves in a certain way, and the constitution of the universe at any point in time is unique. From the observations to date, it would appear that the universe is about ten billion years old, which fits in with the notion that the universe was, at an earlier date, expanding at a greater rate than it is today. This theory is known popularly as the "big bang" theory. It postulates that at the origin of time, the universe was concentrated with infinite density and then catastrophically exploded outward, and that the galaxies were formed from this primordial material.

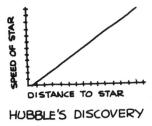

HUBBLE'S DISCOVERY

Competing with this theory is the so-called "steady state" theory, which is more in line with the philosophical thought that the universe endures "from everlasting to everlasting." But the great bulk of the astronomical observations today agrees with the big bang theory rather than the steady-state theory, and the steady state theory has been largely abandoned.

Of course, we are still faced with the unsettling question of what about before the big bang. We do not have the answer. But perhaps the reader will have realized by this point that time has many faces, and perhaps in the long run questions of this sort, relating to the ultimate beginning and end of the universe, are simply projections of our own micro experience into the macroworld of a universe that knows no beginnings and no ends.

## STELLAR CLOCKS

Quite often in science a project that was intended to explore one area stumbles unexpectedly upon interesting results in another. Several years ago, a special radio telescope was built at Cambridge University's Mullard Radio Observatory in England, to study the twinkling of radio stars—stars that emit radio waves. The twinkling can be caused by streams of electrons emitted by the sun. It may be quite fast, so equipment was designed to detect rapid changes.

In August of 1964, a strange effect was noticed on a strip of paper used to record the stellar radio signals: There was a group of sharp pulses bunched tightly together. The effect was observed for over a month and then disappeared, only to reappear. Careful analysis indicated that the pulses were coming with incredible regularity at the rate of 1.33730113 per second, and each pulse lasted 10 to 20 milliseconds. Such a uniform rate caused some observers to suspect that a broadcast by intelligent beings from outer space had been intercepted. But further observations disclosed the presence of other such "stellar clocks" in our own galaxy—the Milky Way Galaxy—and it did not seem reasonable that intelligent life would be so plentiful within our own galaxy.

**PULSAR SIGNAL**

It is generally believed now that the stellar clocks, or *pulsars* as they are called, are neutron stars, which represent one of the

**FORMATION OF STAR**

**STRUGGLE BETWEEN
GRAVITY AND OUTWARD
PRESSURE PRODUCED BY
NUCLEAR FURNACE**

**NUCLEAR FURNACE BURNS
OUT AND STAR COLLAPSES
ON ITSELF**

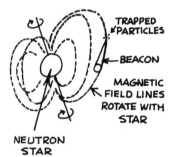

TRAPPED
PARTICLES

BEACON

MAGNETIC
FIELD LINES
ROTATE WITH
STAR

**NEUTRON
STAR**

last stages in the life of a star. According to the theory of the birth, evolution, and death of stars, stars are formed from interstellar dust and gas that may come from debris left over from the initial "big bang" or from the dust of stars that have died in a violent explosion or "super nova."

A particular cloud of gas and dust will begin to condense because of the mutual gravitational attraction between particles. As the particles become more compact and dense, the gravitational forces increase, forming a tighter and tighter ball, which is finally so dense and hot that nuclear reactions, like a continuously exploding H bomb, are set off in the interior of the mass.

### White Dwarfs

In a young star, the energy of heat and light is produced by the nuclear burning of hydrogen into helium. The pressure generated by this process pushes the stellar material outward against the inward force of gravitation. The two forces struggle against each other until a balance is reached. When the hydrogen is exhausted, the star begins to collapse gravitationally upon itself again, until such a high pressure is reached that the helium begins to burn, creating new and heavier elements. Finally, no further burning is possible, no matter what the pressure, and the star begins to collapse under its own weight.

At this point, what happens to the star depends upon its mass. If its mass is near that of our own sun, it collapses into a strange kind of matter that is enormously more dense than matter organized into the materials we are familiar with on earth. One cubic centimeter of such matter weighs about 1000 kilograms. Such a collapsed star is called a "white dwarf," and it shines faintly for billions of years before becoming a "clinker" in space.

### Neutron Stars

For stars that are slightly more massive than our sun, the gravitational collapse goes beyond the white dwarf stage. The gravitational force is so great and the atoms are jammed together so closely that the electrons circling the core of the atom are pressed to the core, joining with the protons to form neutrons with no electrical charge. Normally, neutrons decay into a proton, a massless particle called a neutrino, and a high-speed electron, with a half-life of about 11 minutes—that is, half of the neutrons will decay in 11 minutes. But given the enormous gravitational force inside a collapsed star, the electrons are not able to escape, and thus we have a "neutron star"—a ball about 20 kilometers in diameter, having a density a hundred million times the density of a white dwarf. Such an object could rotate very fast and not fly apart, and it looks as though the neutron star is the answer to the puzzling "stellar clocks."

But where do the pulses come from? Such a neutron star will have a magnetic field that rotates with the star, as the earth's magnetic field rotates with the earth. Electrically charged particles near the star will be swept along by the rotating magnetic field;

and the farther they are from the star, the faster they will have to rotate—like the ice skater on the end of a "crack the whip" chain.

The most distant particles will approach speeds near that of light, but according to the relativity theory no particle can exceed the speed of light. So these particles will radiate energy to "avoid" exceeding the speed of light. If the particles are grouped into bunches, then each time a bunch sweeps by, we will see a burst of light or radio energy as though it were coming from a rotating beacon of light. Thus the pulses we detect on earth are in reality the signals produced as the light sweeps by us. If such an explanation is correct, then the star gradually loses energy because of radio and light emission, and the star will slow down. Careful observations show that pulsar rates *are* slowing down gradually, by an amount predicted by the theory.

### Black Holes—Time Comes to a Stop

Stars with masses about that of our own sun or smaller collapse into white dwarfs; slightly more massive stars collapse into neutron stars. Now let's consider stars that are so massive that they collapse into a point in space.

In the case of the neutron star, complete collapse is prevented by the nuclear forces within the neutrons, but with the more massive stars, gravitation overcomes even the nuclear forces; and according to the theories available today, the star continues to collapse to a point in space containing all of the mass of the original star, but with zero volume, so that the density and gravity are infinite—gravity is so strong near this object that even light cannot escape; hence the term *black hole*.

These fantastic objects—black holes—were postulated theoretically, utilizing relativity theory, in the late 1930's; and within the last few years the evidence is mounting that they do indeed exist. One such observation reveals a star circling around an invisible object in space. In the vicinity of this unseen star, or black hole, strong x-rays are emitted, and it is suspected that these x-rays are generated by matter streaming into the black hole—matter that the gravity field of the black hole pulls away from the companion star.

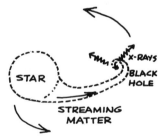

How would time behave in the vicinity of such a strange object? We recall from our section on relativity (page 125) that as the gravitational field increases, clocks run more slowly. Let's apply this idea to a black hole. Suppose we start out with a massive star that has exhausted all fuel for its nuclear furnace and is now beginning to undergo gravitational collapse.

We'll suppose that on the surface of this collapsing star we have an atomic frequency standard whose frequency is communicated to a distant observer by light signals. As the star collapses, the frequency of the atomic standard, as communicated by the light signal, would decrease as the gravitational field increases. Finally, the size of the star reaches a critical value where the gravitational pull is so strong that the light signal is not able to leave the surface of the star.

FORMATION OF BLACK HOLE

Our distant observer would notice two things as the star approaches this critical size: First, the clock on the surface of the star is running more and more slowly; and at the same time, the image of the star is getting weaker. Finally we are left with only the "Cheshire cat smile" of the star.

A careful mathematical analysis of the situation shows that for the distant observer, it appears to take infinite time for the star to reach this critical size; but for an observer riding with the clock on the surface of the star, the critical size is reached in a finite length of time.

What does all of this mean? No one knows for sure. The equations indicate that the massive star just keeps collapsing on itself until it is merely a point in space. Mathematicians call these points in space *singularities;* and when a singularity is encountered in a mathematical law of nature, it means that the theory has broken down and scientists start looking for a more powerful theory that will lead them into new pastures. It has happened many times before in physics. For example, when Niels Bohr postulated that an electron could circle around an atom without spiraling into the nucleus, he provided a stepping-stone toward a whole new concept of the micro-world. Perhaps the "black hole" is the doorway connecting the micro-world to the macro-world.

### TIME, DISTANCE, AND RADIO STARS

In Chapter 12 we described systems for determining distance and location from synchronized radio signals. Here we shall discuss a new technique for relating time to distance via observations of radio stars—a technique that has grown out of the relatively new science of *radio astronomy.*

One of the problems of astronomy is to determine the direction and shapes of distant celestial objects. Astronomers refer to this as the "resolution" problem. The resolution of a telescope is primarily determined by two factors—the area of the device that collects the radiation from outer space, and the radiation frequency at which the observation is made.

RESOLUTION

AREA OF ANTENNA

RADIO FREQUENCY

As we might expect, the bigger the collecting area, the better the resolution; but not so obvious is the fact that resolution decreases as we make observations at lower frequencies. For optical astronomers the area of the collecting device is simply the area of the lens or mirror that intercepts the stellar radiation. And for radio astronomers it is the area of the antenna—quite often in the shape of a dish—that figures in the determination of resolution.

Because of the dependence of resolution on frequency, an optical telescope lens with the same area as a radio telescope dish yields a system with much greater resolution because optical frequencies are much higher than radio frequencies. The cost and engineering difficulties associated with building large radio antennas, to achieve high resolution at radio frequencies, fostered alternative approaches. A system consisting of two small antennas separated by a distance has the same resolution as one large antenna whose diameter is equal to the separation distance. Thus, instead of building one large antenna ten kilometers in diameter, we can achieve the same resolution with two smaller antennas separated by ten kilometers.

But as always, this advantage is obtained at a cost. The cost is that we must very carefully combine the signals received at the two smaller dishes. For large separation distances, the signals at the two antennas are typically recorded on magnetic tape, using high-quality tape recorders.

It is important that the two signals be recorded very accurately with respect to time. This is achieved by placing at the two antenna sites synchronized atomic clocks that generate time signals recorded directly on the two tapes along with the radio signals from the two stars. With the time information recorded directly on the tapes, we can at some later time bring the two tapes together—usually to a location where a large computer is available—and combine the two signals in the time sequence in which they were originally recorded. This is important, for otherwise we will get a combined signal that we cannot easily disentangle.

It is also important that the radio-star signals be recorded with respect to a very stable frequency source; otherwise, the recorded radio-star signals will have variations, as though the radio telescopes were tuned to different frequencies of the radio star "broadcast" during the measurement. The effect would be similar to trying to listen to a radio broadcast while someone else was continually tuning to a new station. The atomic standard also provides this stable frequency-reference signal. These requirements for time and frequency information are so stringent that the two-antenna technique, with large separation distance, is not practical without atomic clocks.

To understand the implications of this technique for synchronization and distance measurement, we need to dig a little more deeply. In the sketch, we see a signal coming from a distant radio star. Signals from radio stars are not at one frequency, but are a jumble of signals at many frequencies; so the signal has the appearance of "noise," as shown in the sketch.

Let's now consider a signal that is just arriving at the two antennas. Because the star is not directly overhead, the signal arriving at antenna A still has an extra distance, $D$, to travel before it reaches antenna B. Let's suppose that it takes the signal a time, $T$, to travel the extra distance, $D$, to antenna B. Thus we are recording the signal at A, a time $T$ before it is recorded at B. The situation is similar to recording a voice transmission from a satellite at two different locations on the earth. Both locations record the same voice transmission, but one transmission lags behind the other in time.

Let's replace the radio star with a satellite. Suppose we know the locations of the satellite and the two earth sites, A and B, as shown in the sketch. We record the two voice transmissions on tape, and later bring the two recordings together and play them

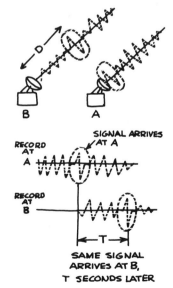

SIGNAL ARRIVES AT A

RECORD AT A

RECORD AT B

T

SAME SIGNAL
ARRIVES AT B,
T SECONDS LATER

back simultaneously. We hear two voices, one being the "echo" of the other.

Now suppose that we have a device that allows us to delay the signal coming out of tape recorder A by an amount that is accurately indicated by a meter attached to the delay device. We adjust the delay from tape recorder A until the two voice signals are synchronized—that is, until the echo has disappeared. The amount of delay required to bring the two voices into synchronism is precisely the delay, $T$, corresponding to the extra distance the signal must travel on its way to antenna B, with respect to antenna A.

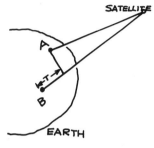

We stated that we knew the locations of the satellite and of A and B. This is enough information to calculate $T$. Suppose $T$ is calculated to be 100 nanoseconds, but that the delay we measure to get rid of the echo is 90 nanoseconds. We are now confronted with a problem. Either the locations of the satellite and of A and B that we used to calculate $T$ are in error, or the atomic clocks at A and B are not synchronized.

We recheck and find that the ground stations and satellite positions are not in error. Therefore, we conclude that the 10 nanoseconds error is due to the fact that the clocks are not synchronized. In fact, they must be out of synchronization by 10 nanoseconds. We now have a new means of synchronizing clocks.

We can also turn this situation around. Suppose we know for certain that the clocks are synchronized, and we also know the position of the satellite accurately. By combining signals recorded at A and B, we can determine what the A-B separation must be to give the measured time lag. Work is now underway to utilize just such techniques—but with radio stars instead of satellites—to measure the distance between distant parts of the surface of the earth to a few centimeters. Such measurements may give new insight into earth crust movements and deformations that may be crucial for the prediction of earthquakes.

The uses to which the relationships of time, frequency, and astronomy may be put are far reaching, and we probably have seen only the beginning.

# Chapter

| | | | | | | |
|---|---|---|---|---|---|---|
| 1 | 2 | 3 | 4 | 5 | 6 | 7 |
| 8 | 9 | 10 | 11 | 12 | 13 | 14 |
| **15** | 16 | 17 | 18 | 19 | 20 | 21 |
| 22 | 23 | 24 | 25 | 26 | 27 | 28 |
| 29 | 30 | 31 | | | | |

# CLOCKWORK AND FEEDBACK

Automation is a cornerstone of modern industrial society. In a sense the clock planted the seed of automation, since its mechanism solves most of the problems associated with building any kind of mechanical device whose sequence of steps is controlled by each preceding step.

A good example is the automatic washing machine. Most such machines have a "timer" that initiates various phases of the wash cycle and also controls the duration of each phase. The timer may "instruct" the tub to fill for 2 minutes, wash for 8 minutes, drain, perform various spray-rinse operations, fill again and rinse, and finally spin-dry for 4 minutes. In most machines the operator can exercise some control over the number and duration of these phases. But unless there is some such interference, once the wash cycle is started, the timer and its associated control components are oblivious to happenings in the outside world.

## OPEN-LOOP SYSTEMS

A control system such as that in an automatic clothes or dish washing machine is called an "open-loop" system, whose main characteristic is that once the process is started, it proceeds through a preestablished pattern at a specified rate. Other examples of devices utilizing open-loop control systems are peanut-vending machines, music boxes, and player pianos. Each such machine is under the control of a clockwork-like mechanism that proceeds merrily along, oblivious to the rest of the world—like the broom in the story, "The Sorcerer's Apprentice," that brings in bucketful after bucketful of water, even though the house is inundated.

## CLOSED-LOOP SYSTEMS

Another important kind of control system, called a "closed-loop" system, employs feedback. An example is a furnace which is activated by a signal from a thermostat. When the room temperature falls below the temperature set on the thermostat, the thermostat produces a signal which is "fed back" to a controlling mechanism at the furnace which turns the furnace on. When the room temperature reaches the value set on the thermostat, the thermostat "instructs" the controlling mechanism to turn the furnace off. This system with its feedback automatically keeps the room at the temperature set on the thermostat. We have already encountered other systems with feedback, or self-regulating systems, in Chapter 2, and again in Chapter 5, in the discussion of atomic clocks.

Systems utilizing feedback are dependent upon time and frequency concepts in a number of ways. We shall explore these in some depth by considering the operation of a radar system that tracks the path of an airplane. Such tracking systems were developed during World War II and were used for the automatic aiming of anti-aircraft guns. Today, tracking radars are used extensively in a variety of ways—such as tracking storms, civilian aircraft, and even bird migrations.

The operational principle of the tracking system is quite simple. A "string" of radar (radio) pulses is transmitted from a radar antenna. If a pulse of radio energy hits an airplane, it is reflected back to the radar antenna, now acting as a *receiving* antenna. This reflected signal, or radar echo, indicates to the radar system the presence of an airplane. If the echo signal strength increases with time, the airplane is moving in toward the center of the radar beam; and if the echo signal strength is *decreasing*, the airplane is moving out of the radar beam.

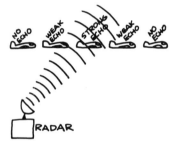

This change of echo signal strength with time is fed back to some device—perhaps a computer—which interprets the echo signal and then "instructs" the radar antenna to point toward the airplane. It all sounds very simple, but, as usual, there are problems.

### The Response Time

The antenna does not respond immediately to changes in the direction of the airplane's flight, for a number of reasons. The inertia associated with the mass of the antenna keeps it from moving at an instant's notice. It also takes time for the computer to interpret the echo signal; and of course there is the delay associated with the travel time of the radar signal itself, to the airplane and back.

These difficulties bring out an important time concept related to feedback systems—namely the "system-response time." Even human beings are subject to this delayed response time, which is typically about 0.3 second. In dinosaurs the problem was particularly serious; a dinosaur 30 meters long would take almost a full second to react to some danger near its tail—if it weren't for an "assistant brain" near the base of its spine!

In our own example, if the system-response time is too long, the plane may move out of the radar beam before the antenna takes a corrective action. The best information in the world is of little use if it is not applied in time.

### System Magnification or Gain

The accurate tracking of an airplane really depends upon the interplay between two factors—the response time just mentioned, and the magnification or *gain* of the feedback system.

We can easily understand this interplay by considering the problem of looking at an airplane through a telescope. At low magnification, or low telescope "gain," the airplane covers only a small portion of the total field of view of the telescope. If the plane takes a sudden turn, we can easily redirect the telescope before the plane disappears from view.

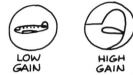

LOW
GAIN

HIGH
GAIN

But with high telescope magnification, the plane will cover a larger portion of the total field of view. In fact, we may be able to see only a portion of the plane, such as the tail section, but in great detail. With high magnification, then, we may not be able to redirect the telescope before the plane disappears from view.

These observations bring us to the conclusion that if we want to track an airplane successfully with a high-magnification or high-gain telescope, we must be able to react quickly. That is, we must have a short *response time*. With lower magnification, we don't have to react so quickly. The obvious advantage of high magnification, from the point of view of tracking, is that we are able to track the airplane with greater accuracy than with low magnification. Or to put it differently, if the telescope is not pointed directly toward the airplane, we won't see it. With lower magnification there is a certain amount of latitude in pointing the telescope while still keeping the plane in view—which means that the telescope may not be pointed squarely at the plane.

These principles of the telescope apply to a radar tracking antenna. The radar radio signal spreads out as a beam from the radar antenna. Depending upon the construction of the antenna, the beam may be narrow or wide, just as a flashlight beam may be narrow or wide. With a narrow beam, all of the radio energy is concentrated and travels in nearly the same direction. If the beam strikes an object, such as the metal surface of an airplane, strong echoes are reflected back to the radar antenna.

On the other hand, we will get no reflections at all from objects in the *vicinity* of the airplane, since they are missed by the narrow radar beam. With a *wide* radar beam, the energy is more dispersed, so we will get only weak reflections, but from objects located throughout a larger volume of space.

WIDE BEAM =
LOW MAGNIFICATION

NARROW BEAM =
HIGH MAGNIFICATION

Thus the *narrow*-beam radar corresponds to the high-magnification telescope, since it gives good information about a small volume of space, whereas the *wide*-beam radar corresponds to the low-magnification telescope, since we get less detailed information about a larger volume of space. With the narrow-beam radar the tracking system must react quickly to changes in direction of the airplane; otherwise the plane may fly out of the beam. And with the wide-beam antenna, more time is available to redirect the antenna before the echoes stop.

Obviously the narrow-beam—high-gain—tracking antenna does a better job of following the path of the airplane, but the price to be paid is that the system must respond quickly to changes in direction of the plane; otherwise, the airplane may be lost from view.

### Recognizing the Signal

There is another kind of difficulty that the radar tracking system may encounter. Not all signals reaching the radar antenna are return echoes from the airplane we are interested in tracking. There may be "noise" from lightning flashes, or reflections from other airplanes, or perhaps even certain kinds of cloud formations. These extraneous signals all serve to confuse the tracking system. If the antenna is to follow the plane accurately, it must utilize only the desired echo signals and screen out and discard all others.

It is at this point that another time and frequency concept, primarily mathematical in nature, comes to our aid. As we shall see, this mathemathical development allows us to dissect the radar signal—or any other signal—into a number of simple components. The dissection gives us insight into its "inner construction," and such information will be invaluable in our task of separating the desired signal from extraneous signals and noise.

### Fourier's "Tinker Toys"

The mathematician most responsible for the development of these ideas was a Frenchman by the name of J. B. J. Fourier, who lived in the early part of the 19th Century. Fourier's development led to a very profound idea—namely, that almost any shape of signal that conveys information can be dissected into a number of simpler signals called *sine waves*. We have already encountered sine waves a number of times in this book, but we haven't called them that. Sine waves are very intimately related to devices that vibrate, or swing back and forth. For example, if we trace out the swinging motion of a pendulum on a moving piece of paper, we have a sine wave.

This sine wave has two important characteristics. First, it has an amplitude indicated by the length of the arrow marked A; and second, it has a pattern that repeats itself once each cycle. The number of cycles per second is the *frequency* of the sine wave; and the length of a particular cycle, in seconds, is the period of the sine wave. In our example there are ten cycles each second, so the frequency is ten cycles per second, or Hz. And the *period* is, therefore, 0.1 second. We can have sets of sine waves all with the same frequency but with differing amplitudes; or sets with differing frequencies, but all with the same amplitude; or sets with differing amplitudes and differing frequencies.

Fourier discovered that with the proper set of sine waves of differing amplitudes and frequencies, he could construct a signal of

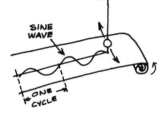

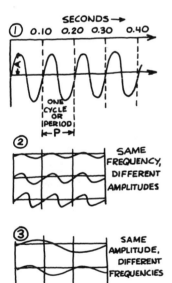

almost any shape. We can think of sine waves as the "tinker toys" out of which we can construct different signals. Let's see how this works.

Suppose we'd like to construct a signal with a square wave shape, like the one shown in the sketch. Since each cycle of the square wave is identical to its neighbors, we need consider only how to build *one* square-wave cycle; all others will be constructed from the same recipe. The sketch shows one cycle of the square wave magnified. The sine wave marked "A" approximates the shape of the square wave, and if we were for some strange reason restricted to only one sine wave in our building of the square wave, this is the one we should pick.

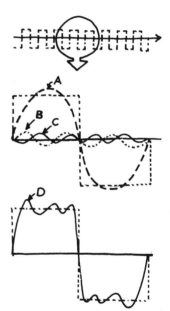

In a sense this particular sine wave represents to the communications engineer what a roughed-out piece of marble represents to the sculptor, and further additions of sine waves represent refinements of the square wave in the same sense that further work on the marble brings out details of the statue. By adding the sine waves B and C to A, we get the signal marked D, which, as we see, is an even closer approximation to the square wave. This process of adding or superimposing sine waves is similar to what happens when ocean waves of different wave lengths come together; the merging ocean waves produce a new wave, whose detailed characteristics depend upon the properties of the original constituent ocean waves.

If we wished, we could add even more sine waves to A, besides B and C, and obtain an even closer approximation to a square wave. Fourier's recipe tells us precisely what sine waves we need to add. We shall not go into the details here, but as a rule of thumb, we can make a general observation: If our signal is very short—such as a pulse of energy one microsecond long—then it takes many sine waves covering a wide range of frequencies to construct the pulse. If, on the other hand, the signal is long and does not change erratically in shape, then we can get by with fewer sine waves covering a narrower range of frequencies.

This concept of relating pulse length to range of frequencies is also the mathematical underpinning for the subject we covered in Chapter 4, decay time—which is one divided by the frequency

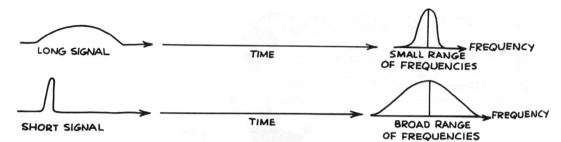

width of the resonance curve. We recall that a pendulum with a long decay time, because of low friction, will respond only to pushes at rates corresponding to a narrow range of frequencies at or near its own natural frequency. In a similar mathematical sense, a radar signal that lasts for a long time—in a sense, that has a long decay time—can be constructed from sine waves covering a narrow range of frequencies. A radio pulse lasting only a short while—corresponding to a pendulum with a short decay time—requires sine waves covering a broad range of frequencies—in the same sense that a pendulum with a short decay time will respond to pushes covering a broad range of frequencies.

### Finding the Signal

Now we must relate Fourier's discovery to the problem of extracting weak radar-echo signals from a noisy background. The problem is somewhat similar to building a cage to trap mice, where the mice play the role of the radar echo signal and the cage is the radar receiver. One of the most obvious things to do is to make the trap door into the cage only big enough to let in mice, and to keep out rats, cats, and dogs. This corresponds to letting in only that range of frequencies necessary to make up the radar signal we are trying to "capture." To let in a wider range of frequencies won't make our signal any stronger, and it may let in more noise—rats and cats—which will only serve to confuse the situation.

But Fourier's discovery suggests how we can build an even better radar receiver that will not only keep out rats and cats, but hamsters too, which are the same size as mice. That is, Fourier tells us how to separate signals of different *shapes*, even though they can be constructed from different "bundles" of sine waves covering the same range of frequencies.

The length of the signal primarily determines the frequency range of the sine waves that we need to build the signal, but within this range we can have many bundles of sine waves to construct many different kinds of signals simply by adding together sine waves with different frequencies, amplitudes, and phases. To go back to the tinker-toy analogy, we can build many different kinds of figures—signals—from tinker toys that are all restricted to a given range of lengths—a certain range of frequencies.

We construct our radar receiver system not only to let in the correct range of frequencies, but also to give favorable treatment to those sine waves that have the amplitudes, frequencies, and phases of exactly the set of sine waves that make up our radar signal; in this way we can separate the mice from the hamsters. Only the very best radar receivers utilize this approach because it is usually expensive; and it requires a high level of electronic circuitry, which relies heavily on time and frequency technology.

To complete the story, there is another approach to separating the mice from the hamsters, which is informationally equivalent to the approach just described, but which is different from an equipment point of view. It is called *correlation detection* of signals, and it simply means that the radar receiver has a "memory," and built into this memory is an image of the signal it is looking for. Thus it can accept signals that have the correct image, and reject those that have not. It is equivalent to the system just described because all of the electronic circuitry required to give favorable treatment to the correct bundle of sine waves is equivalent, from an information point of view, to having an image of the desired signal built into the receiver.

**CORRELATION DETECTION**

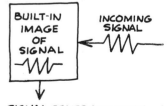

**SIGNAL COMES OUT WHEN RECEIVED SIGNAL AND IMAGE MATCH**

## CHOOSING A CONTROL SYSTEM

We have discussed two kinds of control systems—the open-loop system, which churns along mindlessly in the face of any changes in the outside world; and the closed-loop system, which responds to changes in the outside world. As we have seen, both systems are intimately related in one way or another to time and frequency concepts. But in terms of operation they are almost at opposite ends of the pole.

We might wonder why the open-loop approach is used in some applications and the closed-loop in others. Perhaps the most crucial test relates to our completeness of knowledge of the process or mechanism that we wish to control. The process of washing clothes is straightforward and predictably the same from one time to the next. First wash the clothes in detergent and water, then rinse, then spin-dry. This predictability suggests the simpler, less expensive, open-loop control system, which, as we know, is what is used.

But some processes are very sensitive to outside influences that are not predictable in advance. Driving to work every morning between home and office building is very routine—almost to the point of being programable in advance—but not quite. If an oncoming car swerves into our lane, we immediately appreciate the full utility of closed-loop control because we can, we hope, take action to avoid a head-on collision.

Sometimes the question of closed-loop versus open-loop control reduces to one of simplicity and economy. Thus quite often we elect to use a closed-loop control, even though, in principle, there are no unknowns that might affect the desired goal. All factors, including available technology that could be included as part of a system, must be considered, and costs and other considerations balanced against benefits. Each kind of system has its advantages and limitations. Both depend upon applications of time and frequency information and technology for their operation.

OPEN LOOP OR
CLOSED LOOP?
ECONOMICS
TECHNOLOGY
ADVANTAGES
LIMITATIONS

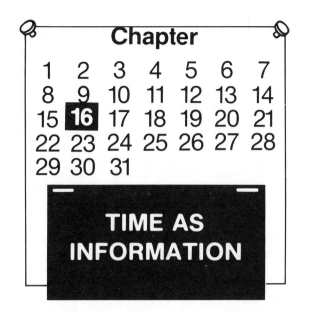

# Chapter

| 1 | 2 | 3 | 4 | 5 | 6 | 7 |
| 8 | 9 | 10 | 11 | 12 | 13 | 14 |
| 15 | **16** | 17 | 18 | 19 | 20 | 21 |
| 22 | 23 | 24 | 25 | 26 | 27 | 28 |
| 29 | 30 | 31 | | | | |

## TIME AS INFORMATION

Many questions in science and technology seek answers to such details as: When did it happen? How long did it take? Did anything else happen at the same time, or perhaps at some related time after? And finally, Where did it happen? We have already seen from the point of view of relativity that questions of when and where have no absolute answers; particularly at speeds approaching the speed of light, the separation between space and time becomes blurred. But in our discussion here we shall assume that speeds are low, so the absolute distinction between space and time—that Newton visualized—holds.

## THREE KINDS OF TIME INFORMATION REVISITED

The question of "when" something happened is identified with the idea of *date*. The question of "how long" it took is identified with time *interval*. And the question of "simultaneous occurrence" is identified with *synchronization*—as we discussed more fully in Chapter 1.

In science, the concept of date is particularly important if we are trying to relate a number of diverse events that may have occurred over a long period of time. For example, we may be taking temperature, pressure, wind speed, and direction measurements at a number of points both upon and above the surface of the earth. For the purpose of weather forecasts, the concept of date is very convenient for such weather measurements because there are a number of persons gathering information at different times of the day, month, and year, scattered over many continents.

Not to have one common scheme for assigning times to measurements is at least a very troublesome bookkeeping problem, and at worst could lead to complete uselessness of the time measurements.

On the other hand, we are quite often content to know simply whether some event occurred simultaneously with another event, or after some regular delay. For example, the fact that a car radio always fades when we drive under a steel overpass suggests some cause-and-effect relationship which, at first glance at least, does not appear to be related to date. Although the fading and the passing under the steel structure occur simultaneously, we notice the same effect at 8:20 on the morning of January 9 that we do at 6:30 in the evening on April 24. An important point is that the amount of time information needed to specify synchronism is generally less than that required to specify date, and we may be able to achieve some economies by realizing the distinction between the two.

Finally, as we have noted before, time *interval* is the most localized and restricted of the three main concepts of time—date, synchronization, and time interval. For example, we are quite often concerned only about controlling the duration of a process. A loaf of bread baked for 45 minutes in the morning is just as good as one baked for 45 minutes at night, and a loaf baked for 3 hours in the morning is just as burned as one baked for 3 hours at night.

To get a better handle on the information content of the three kinds of time, let's return to the problem of boiling an egg. Suppose we have a radio station that broadcasts a time tick once a minute, and nothing else. If we want to boil an egg for three minutes, such a broadcast is quite adequate. We simply drop an egg into boiling water upon hearing a tick, and take it out after three more ticks.

But let's suppose now that a next-door neighbor would like to prepare a three-minute egg, and for some strange reason he would like to boil his egg at the same time we are boiling ours. He can use the radio time signal to make certain that he boils the egg for three minutes, but he can't use the signal to assure himself that he will start his egg when we start ours. He needs some added information. We might arrange to flash our kitchen lights when we start our egg, which will signal him to start his.

But of course this wouldn't be a very practical solution if, for some even stranger reason, everyone in the whole town would like to boil eggs when we boil ours. At this point it would be more practical to include a voice announcement on the radio time signal, which simply says, "When you hear the next tick, drop your egg into the water."

We have now solved the *simultaneity* problem by adding extra information to the broadcast, but even this solution is not altogether satisfactory, for it means that everyone in the whole town must keep his radio turned on all the time waiting for the announcement saying, "Drop your egg into the water." A much more satisfactory arrangement is to announce every hour that all eggs will be dropped into the water on February 13, 1977, at 9:00 A.M.—and to further expand the time broadcasts to include announcements of the date, say every five minutes.

Thus as we progress through the concepts of time interval, simultaneity or synchronism, and date, we see that the information content of the broadcast signal must increase. In a more general sense, we can say that as we move from the localized concept of time interval to the generalized concept of date, we must supply more information to achieve the desired coordination. And as usual, we cannot get something for nothing.

## TIME INFORMATION—SHORT AND LONG

Generally speaking, we associate time information with clocks and watches. Time interval—if the interval of interest is shorter than half an hour or so—is often measured with a stop watch. Where greater precision is required, we can use some sort of electronic time-interval counter. But some kinds of time information are either too long or too short to be measured by conventional means. In our discussion of Time and Astronomy we deduced an age for the universe from a combination of astronomical observations and theory. Obviously, no clock has been around long enough to measure directly such an enormous length of time.

There are also intervals of time that are too short to be measured directly by clocks—even electronic counters. For example, certain elementary atomic particles with names like *mesons* and *muons* may live less than one billionth of a second before they turn into other particles.

If we cannot measure such short times with existing clocks, how do we come to know or speak of such short intervals of time? Again, we *infer* the time from some other measurement that we *can* make. Generally these particles are traveling at speeds near the velocity of light, or about 5 centimeters in a nanosecond ($10^{-9}$ second). When such a particle travels through a material like photographic film, called an emulsion, it leaves a track, the length of which is a measure of the lifetime of the particle. Tracks as short as 5 millionths of a centimeter can be detected, so we infer lifetimes as short as $10^{-15}$ second. But we must restate that we have not actually measured the time directly—we have only inferred it.

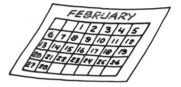

TRACK LENGTH IS A
MEASURE OF PARTICAL
LIFETIME

We can imagine even shorter periods of time, such as the time it would take a light signal to travel across the nucleus of a hydrogen atom—about $10^{-24}$ second. Of course, we can *imagine* even shorter times, such as $10^{-1000}$ second. But no one knows for sure what such short periods of time mean, because no one has measured directly or indirectly such short times, and we are on very uncertain ground if we attempt to extrapolate what happens over intervals of time that we can measure to intervals well beyond our measurement capability.

The question of whether time is continuous or comes in "lumps" like the jerky motion of the second hand on a mechanical watch has occupied philosophers and scientists since the days of the Greeks. Some scientists have speculated that time exists and "passes" in discrete lumps, like energy. (See page 41.) But others think that time is continuous, and that we can divide it into as small pieces as we like, as long as we are clever enough to build a device to do the job, but there is not yet sufficient evidence to decide between the two points of view.

### GEOLOGICAL TIME

We have already seen how cosmological times have been inferred. Here we shall discuss a technique that has shed a good deal of light on the evolution of the earth and life it sustains. Again, as always, we want to tie our time measurements to some mechanism or process that occurs at a regular and predictable rate, and has done so for a very long time. If we want to measure something over a long period of time, we should look for some phenomenon that occurs at a very low rate, so that it won't have consumed itself before our measurement is complete.

One such process is related to carbon 14, which is a radioactive form of carbon. Carbon 14 has a "half-life" of 5000 years. This means that if we took a lump of pure carbon 14 and looked at it 5000 years later, we would find that half of the original lump would still be radioactive, but the rest would have "decayed" to become ordinary carbon. After another 5000 years the half that was radioactive would have been reduced again to half radioactive

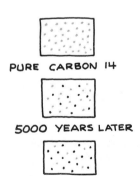

PURE CARBON 14

5000 YEARS LATER

10,000 YEARS LATER

and half non-radioactive carbon. In other words, after 10,000 years the lump would be ¼ radioactive carbon, and ¾ ordinary carbon.

We see that we have a steady process where half of the radioactive carbon present at any particular time will have turned into ordinary carbon after 5000 years. Radioactive carbon 14 is produced by cosmic rays striking the atmosphere of the earth. Some of this carbon 14 will eventually be assimilated by living plants in the process of photosynthesis, and the plants will be eaten by animals. So the carbon 14 eventually finds its way into all living organisms. When the organism dies, no further carbon 14 is taken in, and the residual carbon 14 decays with a half-life of about 5000 years. By measuring the amount of radioactivity, then, it is possible to estimate the elapsed time since the death of the organism, be it plant or animal.

Other substances have different half-lives. For example, a certain kind of uranium has a half-life of about $10^9$ years. In this case the uranium is turning not into non-radioactive uranium, but into lead. By comparing the ratio of lead to uranium in certain rocks, scientists have come to the conclusion that some of these rocks are about five billion years old.

## INTERCHANGING TIME AND LOCATION INFORMATION

As we have said, quite often scientists are interested in *where* something happens, as well as *when* it happens. To go back to our weather measurement, knowing where the weather data were obtained is as important as knowing when they were obtained. The equations that describe the motion of the atmosphere depend upon both location and time information, and an error in either one reduces the quality of weather forecasts.

As a simple illustration, let's suppose that a hurricane is observed as moving 100 kilometers per hour in a northerly direction at 8:00 A.M., and has just passed over a ship 200 kilometers off the coast of Louisiana. Assuming that the storm keeps moving in the same direction at the same speed, it should arrive over New Orleans two hours later, at 10:00 A.M. But the warning forecast could be in error for either or both of two simple reasons: The ship's clock may be wrong, or the ship may be either nearer to or farther from the shore than its navigator thought. In either case, the storm would arrive at New Orleans at a time other than forecast, and the surprised citizens would have no way of knowing which of the two possible errors was responsible for the faulty forecast.

In actual practice, of course, there are other factors that could cause a wrong forecast; the storm might veer off in some other direction, or its rate of movement might slacken or accelerate. But the example illustrates how errors in time or position or both can contribute to faulty predictions; and naturally, the difficulty described here also applies to any process that has both location and time components. Thus we see another kind of interchangeability between time and space that is distinct from the kind that concerned Einstein.

## TIME AS STORED INFORMATION

An important element in the progress of man's understanding of his universe has been his ability to store and transmit information. In primitive societies, information is relayed from one generation to the next through word of mouth and through various ceremonies and celebrations. In more advanced societies, information is stored in books, long-playing records and tape, microfilm, computer memories, and so forth. This information is relayed by radio, television broadcasts, and other communication systems.

We have seen that time is a form of information, but it is also very perishable because of its dynamic nature. It does not "stand still," and therefore cannot be stored in some dusty corner. We must maintain it in some active device, which is generally called a clock. Some clocks do a better job of maintaining time than others. As we have seen, the best atomic clocks would not be in error by more than a second in 370,000 years—whereas other clocks may lose or gain several minutes a day, and may refuse to run altogether after a few years.

Any clock's "memory" of time fades with time, but the rate of the fading differs with the quality of the clock. Radio broadcasts of time information serve to "refresh the clock's memory." We have already hit upon this point in our discussion of communication systems in Chapter 7. We discussed high-speed communication systems in which it is necessary to keep the various clocks in the system synchronized, so that the messages do not get lost or jumbled with other messages. We also stated that quite often the communication system itself is used to keep the clocks synchronized. But the transmission of time to keep clocks synchronized is really a transmission of *information*, so if the clocks in a communication system are of poor quality, a good bit of the information capacity of the system must be used just to keep the clocks synchronized.

A particularly good illustration of this process occurs in the operation of television. The picture on a black-and-white television screen really consists of a large number of horizontal lines that vary in brightness. When viewed from a distance the lines give the illusion of a homogeneous picture.

The television signal is generated at the TV studio by a TV camera, which converts the image of the scene at the studio into a series of short electrical signals—one for each line of the picture displayed on the screen of the TV set. The TV signal also contains information that causes that portion of the TV picture being displayed on the screen to be "locked" to the same portion of the scene at the studio that is being scanned by the TV camera. That is, the signal in the TV set is *synchronized* to the one in the camera at the studio. Thus the TV signal contains not only picture information, but *time* information as well. In fact, a small percentage of the information capacity of a TV signal is utilized for just such timing information.

In principle, if TV sets all contained very good "clocks," which were synchronized to the "clock" in the TV camera at the studio, it would be necessary only occasionally to reset the TV receiver "clock" to the camera "clock." But as a practical matter, such high-quality clocks in TV sets would make them very expensive. So as an alternative there is a rather cheap clock which must be reset with a "synchronization pulse," every 63 microseconds, to keep the clocks running together.

## THE QUALITY OF FREQUENCY AND TIME INFORMATION

There is at present no other physical quantity that can be measured as precisely as frequency. Frequency can be measured with a precision smaller than one part in one hundred thousand billion. Since time interval is the sum total of the periods of many vibrations in the resonator in a clock, it too can be measured with very great precision. Because of these unique qualities of frequency and time among all other physical quantities, the precision—and accuracy—of any kind of measurement can be greatly improved if it can be related in some way to frequency and time.

We have already seen an example of this fact in the operation of navigation systems in which time is converted to distance. Today considerable effort is being devoted to translating measurement of other quantities—such as length, speed, temperature, magnetic field, and voltage—into a frequency measurement. For example, in one device frequency is related to voltage by the "Josephson effect," named for its discoverer Brian Josephson, at the time a British graduate student at Oxford, who shared a 1973 Nobel prize for the discovery. A device called a "Josephson junction," which operates at very low temperatures, can convert a microwave frequency to a voltage. Since frequency can be measured with high accuracy, the voltage produced by the Josephson junction is known with high accuracy; the error is only about two parts in 100 million. This frequency then serves as a reference for voltage.

A related application of this fact is that we may have, someday, a device that can serve as a standard for both length and time. As we have seen, the standard second is based upon a resonant frequency of the cesium atom. The internationally agreed-upon standard of length is no longer based upon a platinum bar, but

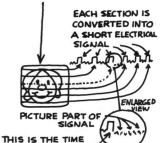

IMAGE AT STUDIO IS "SEPARATED" INTO MANY HORIZONTAL SECTIONS BY TV CAMERA

EACH SECTION IS CONVERTED INTO A SHORT ELECTRICAL SIGNAL

PICTURE PART OF SIGNAL

ENLARGED VIEW

THIS IS THE TIME PART OF SIGNAL WHICH KEEPS THE PICTURE IN THE TV SET IN STEP WITH THE ONE AT THE CAMERA

SIGNAL RECEIVED AT TV SET

THE SHORT ELECTRICAL SIGNALS ARE "REASSEMBLED" AT THE TV SET TO PRODUCE A PICTURE

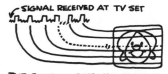

SIGNAL AT MICROWAVE FREQUENCY → JOSEPHSON JUNCTION → VOLTAGE OUT

upon a wavelength of light from the element krypton. The meter is defined as 1,650,763.73 wavelengths of an orange-red light emission that corresponds to a very high frequency—about 50 million times greater than the cesium atom's frequency—which cannot at the present time be measured directly. But considerable work is in progress to establish a link connecting the microwave frequencies that define time to the optical frequencies that define length.

With such a link, either standard could be used to define both a length and a time interval. That is, we would have established a technique for measuring both the wavelength and frequency of the same radiation. Whether such a standard would be derived from signals in the microwave region or at higher frequencies, in the optical region, remains to be seen. Ultimately, the decision will depend on which approach leads to a single standard that is best in absolute accuracy for both frequency and length.

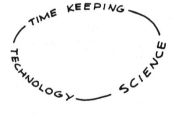

In this and the previous chapters we have been able to mention only a few of the many associations between science and technology on the one hand, and time on the other. It is clear, however, that the progress and advancement of science, technology, and timekeeping are intimately bound together, and at times it is not even possible to make a clear distinction between cause and effect in the advancement of any one of the three. For the most part we have attempted to emphasize those aspects of the development of science, technology, and timekeeping that are clearly established or at least well down the road toward development. In the next and final chapter we shall explore the generation, the dissemination, and the uses of time that lie more in the future than in the present —which, of course, means that we shall be dealing more with speculations than with certainties.

# Chapter

| 1 | 2 | 3 | 4 | 5 | 6 | 7 |
|---|---|---|---|---|---|---|
| 8 | 9 | 10 | 11 | 12 | 13 | 14 |
| 15 | 16 | **17** | 18 | 19 | 20 | 21 |
| 22 | 23 | 24 | 25 | 26 | 27 | 28 |
| 29 | 30 | 31 | | | | |

# THE FUTURE OF TIME

We have called Time "the great organizer." In a world that is rapidly depleting its known natural resources, it is mandatory that we utilize efficiently the resources we have. Central to efficient use are planning, information gathering, organizing, and monitoring. The support of these activities will make great demands on time and frequency technology.

## USING TIME TO INCREASE SPACE

We can think of time and frequency technology as providing a giant grid within which we can file, keep track of, and retrieve information concerning the flow of energy and materials. The higher the level of our time and frequency technology, the more we can pack into the cells of our grid. Improving time and frequency technology means that the walls separating the cells within the grid can be made thinner, thus providing more spacious cells. And at the same time we can identify more rapidly the location of any cell within the system. To explore this theme we shall once again use transportation and communication as examples.

As a safety measure, airplanes are surrounded by a volume of space, into which other planes are forbidden to fly. As the speed of the plane increases, the volume of this space increases proportionately, in much the same way that a driver of a car almost automatically leaves a larger space between himself and other cars on the highway as he increases his speed. Over the years, the average speed and the number of planes in the air have increased dramatically, to the point that there are severe problems in maintaining safety in high-traffic areas.

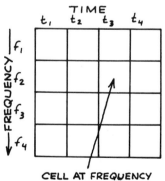

CELL AT FREQUENCY $f_2$ AND TIME $t_3$

We have two choices: Either limit air traffic or institute better air traffic control measures—which, in effect, means reducing the size of the protective space around each plane. At present new systems are being explored that will allow greater airplane density without diminishing safety.

PLANE A TRANSMITTS A PULSE...

...WHICH ARRIVES 5 MICROSECONDS LATER AT PLANE B

One possible system for collision avoidance relies upon the exchange of pulsed radio signals between planes. Participating planes carry synchronized clocks that "time" the transmission of the pulses. For example, plane A might transmit a pulse arriving 5 microseconds later at plane B. As we know, radio signals travel at about 300 meters in one microsecond; so planes A and B are separated by about 1500 meters. This system puts a severe requirement on maintaining interplane synchronization, since each nanosecond of clock synchronization error translates into one foot of interplane error. But with better, more reliable time and frequency technology, the safety factor can be improved.

In Chapter 11 we saw that high message-rate communication systems rely heavily on time and frequency technology so that messages can be both directed to and received at the correct destination. Many of these messages travel in the form of broadcast radio signals, with different kinds of radio "message traffic" being assigned to different parts of the radio frequency spectrum.

Just as a protective space is maintained around airplanes, a protective frequency gap is maintained between radio channels. And further, just as air space is limited, so is radio "space." We cannot use the same piece of radio space for two different purposes at the same time.

To get the best use from the radio space, we would like to pack as much information as possible into each channel, and we would like the protective frequency gap between channels to be as small as possible. Better frequency information means that we can narrow the gap between channels, since there is reduced likelihood of signals assigned to one radio channel drifting over into another. Better time and frequency information together contribute to the possibility of packing more, almost error-free, information into each channel by employing intricate coding schemes.

SIMPLE SIGNAL CARRYING ONLY ONE MESSAGE

SIMPLE RECEIVER

BREAKER, BREAKER
THIS IS BIGFOOT 10—
JEST PASSED OL'
SMOKEY IN A
PLAIN YELLER
WRAPPER—

JOHN ROBB

The transportation and communication examples we have cited can work no better than the underlying technology that supports them. We may be able to generate high-rate messages and build high-speed airplanes by the hundreds, but we cannot launch them into the "air" unless we can assure that they will arrive reliably and safely at their appointed destinations.

In the past the world has operated as though it had almost infinite air space, infinite radio space, infinite energy and raw materials. We are now rapidly approaching the point when the infinite-resource approximations are no longer valid, and our ability to plan and organize will be heavily strained. It is here, no doubt, that time and frequency technology will be one of man's most valuable and useful tools.

**COMPLEX SIGNAL CARRYING MANY MESSAGES**

**COMPLEX RECEIVER TO DETECT AND SEPARATE SIGNAL INTO DIFFERENT MESSAGES**

## TIME AND FREQUENCY INFORMATION—WHOLESALE AND RETAIL

The quality of time and frequency information depends ultimately upon two things—the quality of the clocks that generate the information, and the fidelity of the information channels that disseminate the information. There is not much point in building better clocks if the face of the clock is covered by a muddy glass. In a sense, we might think of the world's standards labs as the wholesalers of time, and the world's standard time and frequency broadcast stations as the primary distribution channels to the users of time at the retail level. Let's explore the possibility of better dissemination systems for the future.

### Time Dissemination

At present the distribution of time and frequency information is a mixed bag. We have broadcasts such as WWV, dedicated primarily to disseminating time and frequency information; and we have navigation signals such as Loran-C, which indirectly provide high-quality time information because the system itself cannot work without it. The advantage of a broadcast such as WWV is that the time information is in a form that is optimized for the users. The signal contains time ticks and voice announcements of time in a readily usable form of information. The formats of navigation signals, on the other hand, are optimized for the purposes of navigation, and time information is in a somewhat buried form, not so easily used.

From the viewpoint of efficient use of the radio spectrum, we would like to have one signal serve as many uses as possible. But such a multi-purpose signal puts greater demands on the user. He must extract from the signal only that information of interest to him, and then translate it into a form that serves his purpose.

In the past, the philosophy has generally been to broadcast information in a form that closely approximates the users' needs, so that processing at the users' end is minimized. This means that the receiving equipment can be relatively simple, and therefore inexpensive. But such an approach is wasteful of the radio spectrum, which is a limited resource. Today, with the development of transistors, large-scale integrated circuits, and mini and micro computers, complicated equipment of great sophistication can be built at a modest cost. This development opens the door to using radio space more efficiently, since the user can now afford the equipment required to extract and mold information to his own needs.

We see that we have a trade-off between receiver complexity and efficient use of the radio spectrum. But there is another aspect of efficient use that we need to explore. The information content itself, of a time information broadcast, is very low compared to most kinds of signals because the signal is so very predictable. A user is not surprised to hear that the time is 12 minutes after the hour when he has just heard a minute before that the time was 11 minutes after the hour. Furthermore, it is important that all

## TIGER

standard time and frequency stations broadcast the same information; we don't want different stations broadcasting different time scales, and causing confusion. In fact, the nations of the world take great pains to see that all stations do broadcast the same time as nearly as possible. But from an information standpoint, the broadcasts are highly redundant.

This redundancy also creeps in, in another way. We don't always readily recognize that there is a good deal of redundant time information in, say, a broadcast from WWV and from a Loran-C station, because the formats of the signals are so vastly different. This redundancy, of course, is not accidental, but intentional—so that, among other reasons, time information can be extracted from Loran-C signals. There are many other examples of systems that carry their own time information in a buried form. We have seen this in the operation of television, where a part of the TV signal keeps the picture in the home receiver synchronized with the scene being scanned by the studio TV camera.

We might ask whether it is necessary to broadcast, in effect, the same time information over and over in so many different systems. Why not have one time signal serve as a time reference for all other systems? Well, there could be advantages in such a plan. But suppose that one universal time and frequency utility serves all, and it momentarily fails. Then all other dependent systems may be in trouble unless they have some backup system.

There is no easy answer to the question of universal time and frequency utility versus many redundant systems. The former is obviously more efficient for radio spectrum space, at the possible costs of escalating failure if the systems falter; whereas the latter is wasteful of radio spectrum space, but it insures greater reliability of operation.

For the immediate future, satellites offer the promise of a vastly improved time signal of the type now offered by WWV. A satellite broadcast could include the same information now provided by WWV—voice announcements of the time, time ticks, standard audio tones, and so forth. And the satellite broadcasts would be enormously superior in terms of accuracy and reliability. A satellite time signal is not subject to the same degree of path delay variation and signal fading that limits shortwave time broadcasts; we could anticipate a thousand-fold increase in accuracy of reception and practically no loss of signal except perhaps for malfunctions of the satellite.

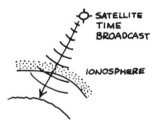

At the present time a radio frequency near 400 MHz has been reserved by international agreement for the broadcast of a satellite time signal, and considerable effort is being made by various national standards labs for the establishment of satellite service. Perhaps in the not-too-distant future we can once again look to the sky for time information, as we did in earlier days when we looked to the sun as our standard.

### Clocks in the Future—The Atom's Inner Metronome

If we reflect for a moment on the history of the development of clocks, we notice a familiar pattern. First, some new approach

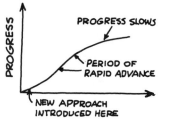

NEW APPROACH
INTRODUCED HERE

**FORCES**
GRAVITATIONAL
ELECTROMAGNETIC
WEAK
NUCLEAR

such as the pendulum—or in more recent times the atomic resonator—is introduced. Because of the intrinsic qualities of the new resonator, a big step forward results. But no resonator is perfect. There is always some problem to be overcome, whether it is compensating for variations in the length of the pendulum caused by temperature fluctuations or reducing frequency perturbations caused by collisions in atomic resonators. As each difficulty is systematically removed, further progress is gained only with greater and greater effort—we have reached a point of diminishing returns. Finally we reach a plateau, and a "leap forward" requires some radical new approach. But this by no means indicates that we are in a position to abandon the past when a new innovation comes along. Usually the new rests on the old. The atomic clock of today incorporates the quartz-crystal oscillator of yesterday, and it may be that tomorrow's clock will incorporate some version of today's atomic clock.

Our ability to build and improve clocks rests ultimately on our understanding of the laws of nature. Nature seems to operate on four basic forces. The earth-sun clock depends upon the force of gravity described within the framework of gravitational theory. The electrons in the atomic clock are under the influence of electric and magnetic forces, which are the subject of electromagnetic theory. These two forces formed the basis of classical physics.

Modern physics recognizes that there are other kinds of forces in nature. A complete understanding of the decay of radioactive elements into other elements requires the introduction of the so-called "weak force." We encountered this force when we employed radioactive dating to determine the age of rocks and of ancient organic material. The fourth and final force, according to our present understanding, is the *nuclear* force, the force that holds the nucleus of the atom together. Many scientists suspect that the four forces are not independent—that there may be some underlying connection, particularly between electromagnetic and weak forces, which will someday yield to a more powerful, universal theory. The technology of time and frequency will no doubt play an important part in unearthing new data required for the construction of such a theory. At the same time, the science of keeping time will benefit from this new, deeper insight into nature.

A technique that both gives insight into nature and points to possible new ways of building even better clocks is based upon a remarkable discovery, in 1958, by the German physicist, R. L. Mossbauer, who later received a Nobel Prize for his work. Mossbauer discovered that under certain conditions the nucleus of an atom emits radiation with extreme frequency stability. The emissions are called *gamma rays*, and they are a high-energy form of electromagnetic wave, just as light is a lower energy form of electromagnetism.

The $Q$ of these gamma ray emissions is over 10 billion, compared to 10 million for the cesium oscillator described in Chapter 5. These high-$Q$ emissions have permitted scientists to check directly the prediction of Einstein—that photons of light are subject to gravitational forces, even though they have no mass. (We encountered this effect when we discussed the operation of clocks located near black holes.)

A photon falling toward the earth gains energy just as a falling rock gains kinetic energy with increasing speed. However, a photon cannot increase its speed, since it is already moving with the speed of light—the highest speed possible, according to relativity. To gain energy the photon must increase its *frequency* because light energy is proportional to frequency. Using high-$Q$ gamma rays, scientists have verified Einstein's prediction, even though the distance the photons traveled was less than 30 meters.

In Chapter 5 we stated that as we go to higher and higher emission frequencies produced by electrons jumping between orbits, the time for spontaneous emission—or natural life time—gets smaller and that it might eventually get so small that it would

be difficult to build a device to measure the radiation. The high-$Q$ gamma rays are not produced by jumping electrons traveling in orbits around the nucleus of the atom. They come from the nucleus of the atom itself. The situation is similar to the radiation produced by jumping electrons in the sense that the nucleus of the atom undergoes an internal rearrangement, releasing gamma rays in the process. But the natural lifetime of these nuclear emissions is much longer than the equivalent atomic emissions at the same frequency. This suggests that nuclear radiations may be candidates for good frequency standards.

But there are two very difficult problems to overcome. First, as we stated in our discussion of the possible combined time-and-length standard, we are just now approaching the point when we can connect microwave frequencies to optical frequencies; and the ability to make a connection to gamma-ray frequencies—some 100 thousand to 20 million times higher in frequency than light—is not close at hand. Second, we must find some way to produce gamma-ray signals in sufficient strength and purity to serve as the basis for a resonant device. We cannot be certain at this time that such a gamma-ray resonator will form the basis for some new definition of the second, but the discussion does point out that there are new avenues to be explored and that the possibility still exists for improving the *Q* of clocks.

## PARTICLES FASTER THAN LIGHT—AN ASIDE

We have touched upon the four basic forces in nature, and upon the theories associated with these forces. Relativity theory says that no object can be made to travel faster than the speed of light. Each small gain in the speed of an object requires greater energy input until, at the speed of light, the energy input is infinite.

But what about the possibility of particles that are already traveling at speeds greater than the speed of light? These particles were not *pushed across* the speed-of-light barrier; they have *always* been on the other side. Such particles have been named *tachyons,* and—if they exist—they must have some remarkable properties if they are to conform to man's present conception of the laws of nature. For example, tachyons gain in speed as they lose energy. Tachyons at rest have an "imaginary" mass—that is a mass which is multiplied by $\sqrt{-1}$. The symbol $\sqrt{-1}$ is well known to mathematicians and can easily be manipulated, but it does not correspond to something that can be measured. This does not present a problem for tachyons, since they are never at rest and so there is no imaginary mass to measure.

But what does all of this have to do with time? When tachyons were first discussed, they seemed to violate—for observers moving in a certain range of speed less than the speed of light—two cornerstones of physics: The law of causality—which says that cause always precedes effect (a timelike idea)—and the idea that something cannot be created out of nothing.

Suppose we have two atoms, A and B. For observers moving in a critical range of speeds it appears as though atom B absorbs a negative-energy tachyon before it is emitted by atom A—a clear violation of causality.

The negative-energy tachyon suggests that we can create particles out of nothing, so to speak. A cardinal rule of physics is that mass/energy is conserved. A system with net zero mass/energy must always have net zero mass/energy. But with negative-energy particles we can create energy out of nothing and not violate conservation of energy. For each new positive energy particle we create, we also create its negative energy cousin, so that, on balance, the net energy is zero.

Happily, a way out of this seeming dilemma was found. Let's suppose that for those observers moving in the critical speed range we interpret the observations differently. We say that instead of seeing a negative-energy tachyon being absorbed by atom B before being emitted by atom A, they see instead a positive-energy tachyon being emitted by B and then absorbed by A.

The search is now on for tachyons, but to date there is no concrete evidence that they exist. For the present, the best we can say

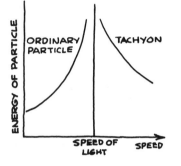

is that tachyons are the product of enlightened scientific imagination.

## TIME SCALES OF THE FUTURE

The history of timekeeping has been the search for systems that keep time with greater and greater uniformity. A point was eventually reached with the development of atomic clocks where time generated by these devices was more uniform than that generated by the movement of the earth around its own axis and around the sun. As we have seen, celestial navigation and agriculture depend for their timekeeping requirements upon the angle and the position of the earth with respect to the sun. But communication systems are not concerned about the position of the sun in the sky. For them, uniform time is the requirement.

As we have pointed out, our present system for keeping time —UTC— is a compromise between these two points of view. But we seem to be moving in the direction of wanting uniform time more than we want sun-earth time. Even navigators are depending more and more on electronic navigation systems. The need for the leap second may be severely challenged at some future date. Perhaps we should let the difference between earth time and atomic time accumulate, and make corrections only every 100 years—or perhaps every 1000 years. After all, we make adjustments of even greater magnitude twice every year, switching back and forth between standard and daylight-saving time. But before we conclude that pure atomic time is just around the corner, we can look back to other attempts to change time and be assured that no revolutions are likely.

### The Question of Labeling—A Second is a Second is a Second

More and more of the world is going to a measurement system based upon 10 and powers of 10. For example 100 centimeters equal a meter, and 1000 meters equal a kilometer. But what about a system where 100 seconds equal an hour, 10 hours equal one day, and so forth? Subunits of the second are already calculated on the decimal system, with milliseconds (0.001 second) and microseconds (0.000001 second), for example. In the other direction we might have the "deciday"—one deciday equals 2.4 hours; the "centiday"—14 minutes and 24 seconds; and the "milliday"—86.4 seconds.

The idea of the decimalized clock is not new. In fact, it was introduced into France in 1793, and, as we might imagine, was met with anything but overwhelming acceptance. The reform lasted less than one year.

Will we someday have decimal time? Possibly. But the answer to this question is more in the realm of politics and psychology—as well as economics—than in technology.

### Time Through the Ages

| | |
|---|---|
| Reality is flux and change. | *"You cannot step twice into the same river, for fresh waters are ever flowing in upon you."* Heraclitus 535–475 B.C. |
| Time involves measure and order. | *Time is the "numerical aspect of motion with respect to its successive parts."* Aristotle 384–322 B.C. |
| Time and space are absolute and separate. | *"Absolute, true and mathematical time, itself and from its own nature, flows without relation to anything external."* Newton 1642-1727 |
| Time and space are relative. | *"There is no absolute relation in space, and no absolute relation in time between two events, but there is an absolute relation in space and time . . ."* Einstein 1879–1955 |

## WHAT IS TIME—REALLY?

We have seen that there are almost as many conceptions of time as there are people who think about it. But what *is* time—really? Einstein pondered this problem when he considered Newton's statements about absolute space and absolute time. The idea of speed—so many miles or kilometers per hour—incorporates both space—distance—and time. If there is absolute space and absolute time, is there then absolute speed—with respect to nothing? We know what it means to say that an automobile is moving at 80 kilometers per hour with respect to the ground; the ground provides a frame of reference. But how can we measure speed with respect to nothing? Yet Newton was implying just this sort of thing when he spoke of absolute space and time.

Einstein recognized this difficulty. Space and time are meaningful only in terms of some frame of reference such as that provided by measuring sticks and clocks, not by empty space. Without such frames of reference, time and space are meaningless concepts. To avoid meaningless concepts, scientists try to define their basic concepts in terms of operations. That is, what we think about time is less important to defining it than how we measure it. The operation may be an experimental measurement, or it may be a statement to the effect that if we want to know how long a second is, we build a machine that adds up so many periods of a certain vibration of the cesium atom.

For scientists, at least, this operational approach to definitions avoids a good deal of confusion and misunderstanding. But if history is our guide, the last word is not yet in. And even if it were, time may still be beyond our firm grasp. In the words of J. B. S. Haldane—

> *"The universe is not only queerer than we imagine,*
> *but it is queerer than we* can *imagine."*

## Some suggestions for further reading

J. Bronowski: *The Ascent of Man* (British Broadcasting Corporation, 1973).

P. C. W. Davies: *The Physics of Time Asymmetry* (University of California Press, 1974).

Donald De Carle: *Horology* (Dover Publications, Inc., 1965).

Richard Feynman: *The Character of Physical Law* (The M.I.T. Press, 1965).

Walter R. Fuchs: *Physics for the Modern Mind* (MacMillan Co., 1967).

Samuel A. Goudsmit and Robert Claiborne: *Time* (Time-Life Books, 1969).

J. B. Priestley: *Man and Time* (Doubleday 1964).

Richard Schlegel: *Time and the Physical World* (Michigan State, 1961).

Albert E. Waugh: *Sundials: Their Theory and Construction* (Dover Publications, Inc., 1973).

G. J. Whitrow: *The Natural Philosophy of Time* (Harper, 1961).

# INDEX

A CATALOG OF SELECTED
**DOVER BOOKS**
IN ALL FIELDS OF INTEREST

# A CATALOG OF SELECTED DOVER BOOKS IN ALL FIELDS OF INTEREST

DRAWINGS OF REMBRANDT, edited by Seymour Slive. Updated Lippmann, Hofstede de Groot edition, with definitive scholarly apparatus. All portraits, biblical sketches, landscapes, nudes. Oriental figures, classical studies, together with selection of work by followers. 550 illustrations. Total of 630pp. 9⅛ × 12¼.
21485-0, 21486-9 Pa., Two-vol. set $29.90

GHOST AND HORROR STORIES OF AMBROSE BIERCE, Ambrose Bierce. 24 tales vividly imagined, strangely prophetic, and decades ahead of their time in technical skill: "The Damned Thing," "An Inhabitant of Carcosa," "The Eyes of the Panther," "Moxon's Master," and 20 more. 199pp. 5⅜ × 8½. 20767-6 Pa. $4.95

ETHICAL WRITINGS OF MAIMONIDES, Maimonides. Most significant ethical works of great medieval sage, newly translated for utmost precision, readability. Laws Concerning Character Traits, Eight Chapters, more. 192pp. 5⅜ × 8½.
24522-5 Pa. $4.50

THE EXPLORATION OF THE COLORADO RIVER AND ITS CANYONS, J. W. Powell. Full text of Powell's 1,000-mile expedition down the fabled Colorado in 1869. Superb account of terrain, geology, vegetation, Indians, famine, mutiny, treacherous rapids, mighty canyons, during exploration of last unknown part of continental U.S. 400pp. 5⅜ × 8½. 20094-9 Pa. $7.95

HISTORY OF PHILOSOPHY, Julián Marías. Clearest one-volume history on the market. Every major philosopher and dozens of others, to Existentialism and later. 505pp. 5⅜ × 8½. 21739-6 Pa. $9.95

ALL ABOUT LIGHTNING, Martin A. Uman. Highly readable non-technical survey of nature and causes of lightning, thunderstorms, ball lightning, St. Elmo's Fire, much more. Illustrated. 192pp. 5⅜ × 8½. 25237-X Pa. $5.95

SAILING ALONE AROUND THE WORLD, Captain Joshua Slocum. First man to sail around the world, alone, in small boat. One of great feats of seamanship told in delightful manner. 67 illustrations. 294pp. 5⅜ × 8½. 20326-3 Pa. $4.95

LETTERS AND NOTES ON THE MANNERS, CUSTOMS AND CONDITIONS OF THE NORTH AMERICAN INDIANS, George Catlin. Classic account of life among Plains Indians: ceremonies, hunt, warfare, etc. 312 plates. 572pp. of text. 6⅛ × 9¼. 22118-0, 22119-9, Pa. Two-vol. set $17.90

ALASKA: The Harriman Expedition, 1899, John Burroughs, John Muir, et al. Informative, engrossing accounts of two-month, 9,000-mile expedition. Native peoples, wildlife, forests, geography, salmon industry, glaciers, more. Profusely illustrated. 240 black-and-white line drawings. 124 black-and-white photographs. 3 maps. Index. 576pp. 5⅜ × 8½. 25109-8 Pa. $11.95

AMERICAN CLIPPER SHIPS: 1833–1858, Octavius T. Howe & Frederick C. Matthews. Fully-illustrated, encyclopedic review of 352 clipper ships from the period of America's greatest maritime supremacy. Introduction. 109 halftones. 5 black-and-white line illustrations. Index. Total of 928pp. 5⅜ × 8½.

25115-2, 25116-0 Pa., Two-vol. set $17.90

TOWARDS A NEW ARCHITECTURE, Le Corbusier. Pioneering manifesto by great architect, near legendary founder of "International School." Technical and aesthetic theories, views on industry, economics, relation of form to function, "mass-production spirit," much more. Profusely illustrated. Unabridged translation of 13th French edition. Introduction by Frederick Etchells. 320pp. 6⅛ × 9¼. (Available in U.S. only)

25023-7 Pa. $8.95

THE BOOK OF KELLS, edited by Blanche Cirker. Inexpensive collection of 32 full-color, full-page plates from the greatest illuminated manuscript of the Middle Ages, painstakingly reproduced from rare facsimile edition. Publisher's Note. Captions. 32pp. 9⅜ × 12¼.

24345-1 Pa. $4.95

BEST SCIENCE FICTION STORIES OF H. G. WELLS, H. G. Wells. Full novel *The Invisible Man*, plus 17 short stories: "The Crystal Egg," "Aepyornis Island," "The Strange Orchid," etc. 303pp. 5⅜ × 8½. (Available in U.S. only)

21531-8 Pa. $6.95

AMERICAN SAILING SHIPS: Their Plans and History, Charles G. Davis. Photos, construction details of schooners, frigates, clippers, other sailcraft of 18th to early 20th centuries—plus entertaining discourse on design, rigging, nautical lore, much more. 137 black-and-white illustrations. 240pp. 6⅛ × 9¼.

24658-2 Pa. $6.95

ENTERTAINING MATHEMATICAL PUZZLES, Martin Gardner. Selection of author's favorite conundrums involving arithmetic, money, speed, etc., with lively commentary. Complete solutions. 112pp. 5⅜ × 8½.    25211-6 Pa. $2.95

THE WILL TO BELIEVE, HUMAN IMMORTALITY, William James. Two books bound together. Effect of irrational on logical, and arguments for human immortality. 402pp. 5⅜ × 8½.    20291-7 Pa. $7.95

THE HAUNTED MONASTERY and THE CHINESE MAZE MURDERS, Robert Van Gulik. 2 full novels by Van Gulik continue adventures of Judge Dee and his companions. An evil Taoist monastery, seemingly supernatural events; overgrown topiary maze that hides strange crimes. Set in 7th-century China. 27 illustrations. 328pp. 5⅜ × 8½.    23502-5 Pa. $6.95

CELEBRATED CASES OF JUDGE DEE (DEE GOONG AN), translated by Robert Van Gulik. Authentic 18th-century Chinese detective novel; Dee and associates solve three interlocked cases. Led to Van Gulik's own stories with same characters. Extensive introduction. 9 illustrations. 237pp. 5⅜ × 8½.

23337-5 Pa. $5.95

*Prices subject to change without notice.*

Available at your book dealer or write for free catalog to Dept. GI, Dover Publications, Inc., 31 East 2nd St., Mineola, N.Y. 11501. Dover publishes more than 175 books each year on science, elementary and advanced mathematics, biology, music, art, literary history, social sciences and other areas.